Whatever Happened to the Metric System?

Anonyponymous
Toponymity
Harold's Tail
The 9 Lives of Alexander Baddenfield
Madeline at the White House

Whatever Happened to the Metric System?

How America Kept Its Feet

John Bemelmans Marciano

BLOOMSBURY

NEW YORK • LONDON • NEW DELHI • SYDNEY

Published by Bloomsbury USA, New York

Bloomsbury is a trademark of Bloomsbury Publishing Plc

All papers used by Bloomsbury USA are natural, recyclable products made
from wood grown in well-managed forests. The manufacturing processes
conform to the environmental regulations of the country of origin.

LIBRARY OF CONGRESS CATALOGING-IN-
PUBLICATION DATA HAS BEEN APPLIED FOR.

ISBN: 978-1-60819-475-9

First U.S. edition 2014

1   3   5   7   9   10   8   6   4   2

Typeset by Hewer Text UK Ltd, Edinburgh
Printed and bound in the U.S.A. by Thomson-Shore Inc, Dexter, Michigan

Bloomsbury books may be purchased for business or promotional use.
For information on bulk purchases please contact Macmillan Corporate
and Premium Sales Department at specialmarkets@macmillan.com.

For Meredith, who loved the metric system

CONTENTS

$1/16$

THE DAY THE METRIC DIED

T HE REAGAN budget, atrocities in El Salvador, the Polish Solidarity movement, Three Mile Island—these were some of the headlining stories on the *CBS Evening News* broadcast of February 19, 1982. After getting through all these reports and more, a teaser tried to get viewers to stick through another slate of commercials for the final segment of the evening, "A Defeated Measure."

The story appeared a few minutes before 7:00 P.M. EST and was introduced by Dan Rather, the anchor who had replaced Walter Cronkite the year before. "Other countries have done it for years. But in this country conversion to the metric system didn't draw many converts and only inched along," Rather said, and then kicked the segment to the reporter. "Bob Schieffer has more on what it takes to kill a meter."

"After a lot of hoopla and millions of tax dollars spent on commercials and other things to tell us why we needed to go metric, it is the metric system that's about to go, a new victim of the Reagan budget ax."

What Schieffer was talking about was not really the metric system itself, but the U.S. Metric Board, an organization put in place by the 1975 Metric Conversion Act that was meant to guide America through the process of abandoning its customary measures for metric ones. Now, however, it was the board itself that was being abandoned.

"In California's Silicon Valley," Schieffer continued, "they love metrics, a system developed by scientists for scientists." Just such a person then came on to explain why the metric system was "simpler, more logical, and probably more accurate."

After listing a few products sold metrically in the United States—wine, liquor, skis, and cameras—Schieffer said that the meter had never really caught on, and that perhaps it was because it was more suited to technology, "and our way more suited to humans."

To explain this line of thinking, on came Stewart Brand, a Northern Californian who would shortly coin the phrase "information wants to be free." His greatest accomplishment was the *Whole Earth Catalog*, what Steve Jobs would later call "Google in paperback form, thirty-five years before Google came along."

Brand took the broadcast viewer on a mini-tour of the measuring capabilities of the human body, as enshrined in America's system of customary measures. He noted that everyone carries one ruler at the end of each leg—their foot—and that the yard was in fact a *reach*, as it was the length of a person's outstretched arm to the center of one's body. Double the yard was the fathom, a handy unit for someone hauling in a line attached to an anchor. By measuring the rope against the length of one's arm span, a person could tell how many fathoms deep the water was.

The *Evening News* didn't put Brand onscreen just because he had a passing interest in anthropomorphic measures; the libertarian proto-techie was one of the leaders of the antimetric movement in America, having even been given a government voice when Governor Jerry Brown appointed him to the California Metric Conversion Council, which Schieffer equated to putting a "fox on chicken coop guard duty."

The next person to pile on against the metric system was Tom Wolfe, whose opposition was also well established. A literary icon in a white suit, Wolfe was the paragon of the New Journalism and author of *The Right Stuff*, a bestseller about the fearless test pilots who became the nation's first astronauts.

Wolfe started by noting that NASA had gone to the moon on inches and pounds and had never considered using any other system. He pointed to intellectuals and people in the arts as being the ones who believed that things were done better in Europe, noting "we're still the most obedient little colonials when it comes to things intellectual."

The report ended with what a few people on the street thought of the metric system: "Too damn confusing"; "I like what we have right now"; and "You can't expect the customer to do homework before shopping."

After being bidden good night by Dan Rather, CBS affiliates switched to local programming, and most of the country watched contestants vie to win a Chevette on *Tic Tac Dough* or Richard Dawson kiss housewives on *Family Feud* and promptly forgot that they were ever supposed to have used kilometers and liters at all.

The Lost Metric Generation

"A Defeated Measure" aired on my twelfth birthday. I, along with most other Americans my age, had grown up being told in grammar school that the United States was going metric. Not that the country was thinking about it, or testing it out, but that it was *going* to happen. There was little reason to bother with the old measures, as they were going to disappear.

America's failed metric experiment had run pretty much concurrently with my own existence. I was still in the crib when a report from the National Bureau of Standards stated it as imperative that the country go metric within ten years, and a couple of months into kindergarten when President Gerald Ford signed the Metric Conversion Act, confirming it would happen. Or so it seemed.

The fact that metrication wasn't mandatory surprised the majority of Americans, and irritated those on the U.S. Metric Board who acted as though it were. Certainly, the message given to the public in general by the government, press, and educators had been that the metric system was going to be our sole system of measures.

For older generations, America's failed switch to the metric system was a blip on the screen or an unloved artifact of the era, like New Coke or the Yugo. Those born in the eighties and beyond, on the other hand, don't realize it ever happened at all. Then there is my generation, which not only remembers it but retains that deep-down feeling that the meter was supposed to have been here by now.

My actual memories of metrication are few. I had a fat blue plastic ruler that fascinated me with its thick lawn of millimeter marks, which seemed so different from the picket-fence-like notches marking the fractions between inches on the foot-long wooden rulers we had lying around the house. I had a coloring book called *Metric Man*, about a superhero so dull and didactic he seemed to have sprung from the id of Jimmy Carter himself. Then there were the two-liter bottles of soda, which seemed remarkable when they came out, being both so huge and the nonglass material so light and unbreakable.

Taken together, these scattered items molded my view of the metric system as new, plastic, bloated, and meant to be virtuous, but boring. It was also something to blame—at least on the farm where I grew up.

One of my jobs as a kid was handing wrenches to my father and older brothers. On any given bolt or nut, the half-inch wrench was a good bet, and if that was too small, the next to try was the five-eighths. If that in turn was too big, maybe we needed a nine-sixteenths. If nothing fit—a frustrating occurrence that happened often enough—we'd curse and say it had to be metric.

Our family business was horses. Like any insular community, the horse world comes with its own specialized vocabulary, but it is particular in preserving so many dialectical relics. Riding pants are called "britches," an old variant of "breeches," a fashion that began to disappear when it became a symbol of aristocrats during the French Revolution. A horse's "withers" are where a harness is buckled, and the word relates to the Old English sense of "with," which meant opposite. "Jockey" and "groom" at one time were both synonyms for "boy," which is why the latter term came to be paired with "bride."

The terminology of horses also includes a couple of otherwise obsolete but once widely used measures, the *furlong* and the *hand*. Races under a mile are held in furlongs, of which there are eight to a mile, while the height of horses is measured in four-inch hands. The cutoff between a horse and a pony comes at 14.2 hands (which is not fourteen and two tenths but fourteen and a half) as measured to the withers. These terms are the only ones used; to call a six-furlong race three quarters of a mile would sound odd, and to say a pony has to be four feet ten inches tall or shorter absolutely ridiculous.

For me, the hand and furlong were measures that meant something, and I could visualize them in a way I couldn't a kilometer or anything else metric, save liters of soda. Not that I thought much about the metric system after I finished school. At least not until I had to.

I moved to Rome in 2000 and spent most of my time learning Italian. In order to make dinner, I also had to learn to talk metric, as nearly everything in the market is bought by the *etto*, which is short for *ettogrammo*, or hectogram. But measures are a lot harder to learn than most foreign vocabulary. Whereas a *casa* is the same thing as a house and a *macchina* precisely a car, an *etto* is about halfway between three ounces and a quarter of a pound. Our standards—feet, pounds, quarts, degrees—are nouns, which we conceive of as something concrete. To think of them as anything different takes a serious taxing of the brain. To convert in my head, I resorted to such mental gymnastics as taking degrees Celsius and doubling them, subtracting a tenth, and adding 32 to get Fahrenheit, a formula that on a 10° Italian day breaks down to

$$10°C \times 2 = 20° - 2 = 18° + 32 = 50°F$$

But such tricks never got me to where I could *feel* Celsius.

Americans in Europe are constantly being called upon to defend their country against all sorts of attacks. *Why do you Americans think you should be different? Why can't you admit when someone else's way is better?* Europeans find our system of measurement a perfect example of our stubborn stupidity. Why on earth do we insist on keeping such a nonsensical, archaic system of measures when there is another system that makes perfect sense and is used by the entire rest of the world?

In answer to such questions, I at times acted like Wolfe's "good little colonial," but I did think that Europeans do certain things better than Americans. In my heart of hearts, however, I never believed that one of them was the metric system.

Body Language

It is difficult to win an argument for America's customary measures on the basis of logic. Such simple questions as *How many feet are in a mile?* or *What's the number of square feet in an acre?* produce the not-so-simple answers of 5,280 and 43,560. Questions that appear to beg even basic logic can be elusive. *What weighs more, a pound of feathers or a pound of gold?* sounds like a trick question for third-graders, for which the answer should be, *A pound is the same no matter what it weighs!* Except it doesn't. A pound of feathers is heavier than one of gold by 22 percent, because plumage and precious metals in the United States are weighed according to different pounds.

With the metric system, such confusion is impossible. There is one name for one measure, one measure for each category, and all ratios are decimal. The simplicity of the metric system is such that most Americans grasp its math better than they do their own system's, if not always consciously. After all, few know that *K*, our abbreviation for a thousand, comes from *kilo*.

Our system is so confused we don't even know what to call it. Many Americans think we use the British imperial measures, which we don't. (If you've ever had a pint in an English pub and gotten drunker than you expected, there are an extra three or so ounces of reasons for it.) "Customary" is the broad term used for nonmetric measures, and ours are generally called the English customary measures, which is odd considering that English people no longer use them.

Despite all this, the vast majority of Americans prefer their system of measurement to metric, and many think it is downright better. But does this preference have any basis other than that it was what we were brought up with?

The interesting thing about the 1970s antimetric movement is that it was not just made up of jingoistic patriots and xenophobes. Tom Wolfe and Stewart Brand didn't fit that description, nor did Isamu Noguchi and William S. Burroughs. Much of their support for customary measures came from their having been derived from man, as Brand demonstrated on TV.

Feet, yards, and fathoms are just the tip of the iceberg as far as anthropomorphic measures are concerned, however. Our body can be turned into a Swiss Army knife of rulers, and throughout history it has been.

Whereas the foot is the measure most of us think of as being body-based, it was the arm that traditionally provided people with the most useful measuring stick. Beyond the fathom equaling our wingspan and the yard being half that, there is the cubit, which is the measure from the tip of the middle finger to the elbow. The cubit is the most storied measure in history, having been used by the pharaohs to build the Great Pyramid and Noah for his ark; via Islam, the measure spread across Africa and Asia as far as Indonesia.

The most varied tool of measurement was the hand. Fingers and thumbs were used by the Romans to divide their foot two different ways, with the finger being a sixteenth of one and a thumb a twelfth. Whereas the English called the latter measure the *inch*, most European languages just substituted their own words for thumb (e.g., *duim* in Dutch).

The most popular hand measure of all was the *span*, which could be contrived any number of ways but most commonly meant the width of a splayed hand from the tip of the thumb to the tip of the pinkie. Our word "span" comes from the hand measure originally, as "fathom" in the sense of "to understand" comes from taking depth measures, both attesting to how prevalent a practice body measuring was.

People had to get creative when it came to reckoning longer distances. The Welsh used a measure called the *leap*, while the English had the *bowshot* (the distance an arrow travels) and the French the *houpée*, which is the range of earshot, or how far a person down the road could cry *Houpée!* and have someone else hear it. Even today we use sound to measure distance, by counting the number of Mississippis between a flash of lightning and the clap of thunder. (Each Mississippi is about a fifth of a mile, no matter what you were told as a kid.)

Such measures lend a touch of the absurd to customary measures, and stories like the yard having been based on a king's girth have always been brought up to discredit the old measures by metric advocates. But nearly all such tales are urban legends, including, one assumes, the belief of the ancient Greeks that their foot was based on that of Hercules and that their *stadion*—the length of the original Olympic track and the standard Greek measure for long distances—was how far the strongman demigod could run while holding his breath.

Beyond body measures, however, a key argument made by many anti-metric Americans—especially architects and artists—was that embedded in the customary measures lay classical systems of natural and harmonious proportions. The metric system is all about decimals, which are easy on the modern brain but not so much when it comes to physically dividing things. Pizzas don't get cut into five or ten slices for a reason. Far from natural or classical, the metric system is a product of the heightened rationalism that marked the Enlightenment.

Debating the merits of different measuring systems is pretty much beside the point, however. The near universality of the metric system has little to do with its being superior and a lot to do with practical and political choices made by nations bent on reinvention. As for America, the argument over the meter doesn't date to the 1970s, but the 1790s. But even earlier, America earned a special place in the history of decimal measures by creating the first one—the U.S. dollar. And the man to whom we owe that distinction, as much as anyone, is Thomas Jefferson.

2/16

or One Eighth

THOMAS JEFFERSON PLANS

THE MILES were racking up on the carriage of Thomas Jefferson as it rolled past the late October foliage on its way north. When he had been appointed to the Congress of the Confederation back in June, his trip from Virginia was supposed to have been a good bit shorter, but Jefferson's course had changed owing to a summer revolt that forced the Congress to bolt the City of Brotherly Love.

The Philadelphia Mutiny of 1783 saw nearly a hundred veterans of the Revolution descend upon what is now Independence Hall demanding back pay. With the state governor unwilling to call out the militia, the fledgling nation's lawmakers fled Philadelphia in advance of marching mutineers on the night of June 24. Relocating to the little town of Princeton, the Congress began meeting in Nassau Hall, the principal building of the local college.

By the time Jefferson arrived at this makeshift capital, he had been traveling for better than two weeks. He made it just in time for the new session, but the first day of the Congress Jefferson attended would also be the last one to meet in New Jersey. The legislature was moving again, this time to Annapolis, right back the way Jefferson had come. It was the Virginian's luck of late.

Thomas Jefferson was coming off a pair of *anni horribiles*. The bad times began in the final days of his none-too-popular stint as wartime governor of Virginia, when in June 1781 Jefferson humiliatingly had to abandon Monticello before advancing British troops, whose progress many believed was the result of Jefferson's neglect of the state militia. An inquiry was launched into the ex-governor's behavior, and while he awaited its results, Jefferson had to turn down the plum job of traveling to Paris to help negotiate an end to the war alongside Benjamin Franklin and John Adams, the friends with whom he had hashed out the Declaration of Independence. He also fell off a horse and broke his arm.

The state exonerated Jefferson of misdoing but did demand that he give back its set of Diderot's *Encyclopédie*, which he did unhappily, the twenty-eight volumes being the greatest collection of knowledge ever assembled in printed form and the ultimate in contemporary Enlightenment thinking.

It was shortly after these setbacks that tragedy of the personal kind struck. The birth of a daughter in May 1782 caused his wife Martha's already fragile health to fail, and she died that fall. Jefferson was so inconsolable that his friends wondered what was wrong with him; his wailing, fainting, and locking himself in a room for three weeks in grief was unusual behavior for a man in 1780s Virginia, no matter how beloved a wife he may have lost.

An opportunity for a change of scenery was offered in November when he was again asked to join the Paris peace talks. Jefferson put his affairs in order and traveled to Philadelphia to catch a France-bound ship, only to be held back by winter weather. With his ship icebound in the harbor, the news arrived that the peace had been concluded, and he trudged back home.

It wouldn't have been unfair to think that Thomas Jefferson's best political days were behind him as his carriage rumbled and shook over the dirt roads south to the Maryland State House in the waning weeks of 1783. At Annapolis, he would be joining a Continental Congress that— beyond being a traveling circus—was all but bereft of the country's best talent. Washington had just said farewell to his troops and would soon be heading home to Mount Vernon, while his "son" Lafayette was already back in Paris, where Adams and Franklin remained to negotiate further treaties. Those in the Congress didn't want to stay; a disgusted Alexander Hamilton had quit to resume his legal career, while others, including Jefferson's political ally James Madison, had taken posts as governors and legislators in their home states, where under the Articles of Confederation the true power lay. The literally drifting Congress was so little regarded even by those who did stick around that it perpetually had trouble getting a quorum.

While the central government had purposely been given little to do, there were two areas of pressing need that had to be addressed by the delegates meeting in Annapolis, both of which Jefferson sank his teeth into.

One concerned a problem that had bedeviled Americans since the days of Plymouth and Jamestown: They had no money. Not that the United States was broke—although that was all too true as well—but the young nation had neither a currency to call its own nor enough coins of anyone else's to go around, and way too much paper. Continentals, the bills of credit that had financed the Revolution with the promised backing of silver, had ceased to circulate, and few who held them ever expected to be repaid at anything close to face value.

The lack of hard currency had long been a major bone of contention with the British, who—preoccupied with their own lack of ready coin— had prohibited their shillings and pence from circulating in the colonies and banned Americans from minting their own. During the Revolutionary War the money shortage got so bad that Indian corn became currency, as wampum shells and the pound of tobacco had been in the early days of the colonies. Small change had never stopped being a problem, and Americans used nails and musket balls to complete transactions.

What coins were getting passed around in the 1780s had been minted by the kings of continental Europe and carried names like *half-joes*, *doubloons, pistoles, pistareens,* and *rix dollars.* The longtime favorite of colonists had been *Dutch dog dollars,* handled so much that the lion emblem had come to look canine, but they had been supplanted by another kind of silver coin. Minted mainly in Mexico, the *peso de ocho* had by the 1780s become the de facto currency of the United States, where it was called the *Spanish dollar* or *piece of eight.* The "eight" referred to *reales,* the Spanish word for "royals" and a lower denomination coin, but in America these eighth parts were called *bits,* which the dollar would often get literally sawed into.

The situation was ridiculous, all the more so because no matter what coin was being used, Americans kept accounts in pounds, shillings, and pence, which also meant using its ancient ratios of 12 pennies to the shilling and 20 shillings to the pound.

Add into this stew the money that the individual states themselves were producing, and you had chaos. Almanacs had grown fat with conversion tables to help Americans keep track of it all. What America needed was a single national currency. On this one topic, most could agree. A coinage plan had already been put to the Congress by Robert Morris. The country's superintendent of finance, the rotund Morris was a voracious speculator and America's foremost money man. The plan, however, had been created by Morris's assistant finance minister, also surnamed Morris but blessed with the unusual Huguenot first name of Gouverneur. Gouverneur was a young man on the rise, a financial prodigy who was the superintendent's protégé. He was also, in the parlance of the day, a rake, whose shenanigans amorous and otherwise had been little slowed by a recent carriage accident that had cost him half of his left leg.

Gouverneur Morris's plan was intricate. It set a base unit of money equal to 1/1140th of a Spanish dollar, capable of encompassing nearly all of the many currencies in use and thus allowing existing ledgers to be squared with a new American monetary standard.

Jefferson offered a simpler plan: Adopt the Spanish dollar. But with a twist. Rather than splitting it into eighths and sixteenths or twelfths and twentieths, Jefferson wanted to take the radical step of dividing the coin by tenths, hundredths, and thousandths—decimal fractions. It was a thing no other nation in the world had ever entirely achieved, not with coins or any other measure.

The Gift of Midas

What does the dollar have to do with the metric system? Or any system of weights and measures? Today, pretty much nothing. But up until recently—and we're talking walking-on-the-moon recently—money was all based on weight. In the late eighteenth century, coinage was not only a part of weights and measures, it was the most vital part, and had been for thousands of years.

Coins and weights were so identified with each other that they could be confused, which is still the case when an Englishman says he is carrying a few extra pounds on him, or with shekels and talents in the Bible. In the Old Testament, these terms refer to weight, but in the New Testament they mean coins. In between the goings-on of the Books of Moses and the writing of the Gospel, a technological revolution occurred.

It happened near the land of King Midas, whose fabled touch was said to account for the rich deposits of gold found in Lydia, a part of present-day Turkey. The brilliance of the coinage idea lay in prepackaging precious metal to make it more easily exchangeable. The Lydians measured out set amounts of the stuff and stamped it, guaranteeing its purity. Their coins—and the concept itself—spread like wildfire across the ancient world.

At first, a word like "shekel" was used interchangeably to mean both weight and coin. Over time, however, coins meant to equal a shekel (or whatever other weight was in use) became lighter. This happened through natural wear, via underhanded practices like clipping and sweating, and most especially on account of unscrupulous kings and emperors, who would mint coins with ever less silver and gold than face value.

The latter-day leaders of Rome were particularly devoted practitioners of debasement, to the point where money ceased to mean much of anything at all. Pepin the Short restored a viable currency to Europe, which his son Charlemagne spread far and wide. The Carolingian *libra* eventually made its way to England, giving the pound its symbols £ and lb. as well as its division into 20s (*solidus* or shilling) and 240d (*denarius* or penny), a none-too-intuitive arrangement later jokingly called the £sd system. It was part of a complete measurement makeover, and the Carolingian pound was in theory the same whether it was made up of silver coins or oats. Unfortunately, French monarchs proved hardly less avid debasers than decadent Roman emperors.

Through the Middle Ages, precious metal was scarce. Europe's economy functioned through feudal dues paid in kind and a few important if hard to come by Mediterranean coins like the *florin* and *ducat*, an equation upended by the discovery of America. The metal that came flooding across the seas from silver mines in South America and Mexico changed everything. Suddenly, European monarchs could obtain the raw material to run vastly expanded mints, which cranked out large silver coins called *thalers*, *daalders*, and *dalers*, or—in English—*dollars*. The Spanish *peso* was the most popular dollar of all, a near-universal currency used not only in Europe and the Americas but across the Pacific, where huge Manila galleons transported cargo loads of them for trade.

Now it was time for the United States to have its own mint, and Thomas Jefferson wanted to make sure that the hand at the press did not belong to Robert Morris. Jefferson didn't care for the way men like the Morrises made their money, as he considered merchants and bankers to be parasites on honest folk like farmers and artisans. Furthermore, the Virginian didn't like Gouverneur on a personal level, although the young New Yorker possessed a lively mind and had created a coinage plan that was in many ways excellent.

Jefferson, however, found the Morris plan to be complicated and too clever by half. "Ingenious" was how he facetiously put it. At first glance it seems that what Jefferson proposed was more conservative, in that he wanted to copy the existing Spanish dollar whereas Gouverneur Morris had devised an all-new American currency. But the raison d'être of the Morris plan was that it should provide continuity; it carried with it the legacy of all the different coins getting passed around, which the New Yorker made work by setting such a tiny base unit it could be multiplied into anything. It was *so* tiny it would be based on a quarter grain of silver, or 1/28,000th of a pound, whereas Jefferson desired a dollar that weighed roughly an ounce.

While Gouverneur Morris wanted to make sure old account ledgers could still be made sense of, Jefferson didn't care a whit about making life easy on bookkeepers. He instead considered that the average citizen was already familiar with the Spanish dollar and understood its worth. By comparison, people would be lost with the new Morris unit, which was unwieldy besides. A horse costing eighty Spanish dollars would sell for 115,200 units under the Morris plan.

The same charges of unfamiliarity and sowing confusion could be leveled at Jefferson's plan, however, with its notion of dividing the dollar by tens. Interestingly, Gouverneur Morris also wanted to employ decimals, but in a different way. He would rid the nation of the eighths, twelfths, sixteenths, and twentieths of British and Spanish coins by virtue of the minuscule nature of his unit, which had a base value so low it couldn't be divided any further, thereby forcing people to use tens to count up. Fractions would be avoided altogether.

The difference between the two methods—tallying up by tens, hundreds, and thousands and dividing down by those same numbers—may seem negligible, but it is critical in the math of measurement. The joining of the decimal tally and the decimal fraction is the great unsung story of numbers, and it transformed how the world was measured.

Disme

The decimal point is so a part of our everyday life that it hardly seems like something that needed inventing. In fact, the decimal fraction was invented lots of times, but it never caught on until after the publication in 1585 of a little volume entitled *Disme*, written by Simon Stevin of Bruges. Stevin would go on to be known as the Dutch Archimedes for his brilliance in mathematics and engineering as well as his fantastical creation the *Zeilwagen* ("sail wagon"), a wind-powered vehicle he built for the Prince of Orange. *Disme*, however, was written while Stevin was still a student at the recently opened University of Leiden, which would become one of Europe's greatest centers of learning.

In its original English translation, Stevin's work was titled *Disme: The Art of Tenths, OR, Decimall Arithmetike, Teaching how to performe all Computations whatsoever, by whole Numbers without Fractions*. It's this last part—numbers without fractions—that was the incredible claim of his Art of Tenths. If every fraction were converted to a base of 10, Stevin maintained, they could be treated like any other Arabic numeral. In fact, whole numbers and fractions could be united in a single expression.

Stevin filled *Disme* with concrete examples of how the Art of Tenths could be used with great ease and power in a variety of fields, from engineering to surveying. The book had the evangelical fire of someone who had stumbled onto a better way—a much better way—and wanted to shout it from the rooftops.

An easy sell, however, it was not. At the time, Hindu-Arabic numerals themselves were only just eclipsing Roman ones, as performing calculations on paper was finally overtaking the counting board. The counting board held one huge advantage, namely, it dealt effortlessly with different numeral bases such as the vigesimal (base-20) shilling or *sou* and the duodecimal (base-12) penny or *denier*. Though first introduced in the West around A.D. 1000, Hindu-Arabic numerals had never been fully grasped nor trusted, and not just in Europe; they had never been widely used in Arab lands or the Indian subcontinent either. And they certainly were not used by the illiterate and innumerate majority of the population in Stevin's day, who etched folk numbers into tally sticks to keep accounts, a method also employed by England's Exchequer.

It was not until the decades following Stevin's death in 1620 that his book began to catch on, and even then only in advanced mathematical circles. The reason for this lies partly in just how radical and artificial a concept the decimal fraction is. Consider the most basic of fractions, the half. The figure "1/2" represents one out of two, literally. To think of a half as .5 requires imagining five parts out of ten to represent what physically is a single entity. A quarter is yet more difficult—twenty-five out of a hundred parts—and downright bizarre is a third, a string of threes out of tens stretching to infinity. And that's just looking at the most common examples.

Dealing with fractions is unavoidable in transactions, and treating them as whole numbers was something people had always contrived to do, with varying solutions. The Roman trick was to base everything on twelfths, which they called *unciae*, the origin of our word "inch." Under this system, basic fractions were whole numbers, with 1/2 equaling six *unciae*, 1/3 four *unciae*, and 1/4 three *unciae*.

For higher math, though, the Romans used the ancient Babylonian fractions, just like modern-day clocks, on which a quarter, third, and half hour become 15, 20, and 30 minutes. Another surviving Babylonian practice is the division of circles into 360 degrees, the ultimate in creating whole numbers out of fractional parts. (Converted to decimals, one degree equals 0.002777.)

Medieval math was far simpler. Everything was based on two. Even advanced multiplying and dividing could be achieved (if slowly) by halving and doubling, again and again and again, leading to fourths and eighths and sixteenths. This is how an ounce—a term that similarly comes from the Latin *uncia*—got to be a sixteenth of a pound or pint instead of a twelfth.

This style of reckoning—sometimes called "octal math"—remained the popular method when Jefferson and the Morrises were presenting their plans, two hundred years after *Disme*'s publication. Decimal fractions remained so cutting-edge that most people alive had never come across them; tally sticks were still in use by the Exchequer. But the educated elite used decimal points, and they had caught Stevin's fervor.

"Every one remembers," Jefferson wrote, "that when learning money arithmetic, he used to be puzzled with adding the farthings, taking out the fours and carrying them on; adding the pence, taking out the twelves and carrying them on; adding the shillings, taking out the twenties and carrying them on; but when he came to the pounds, where he had only tens to carry forward, it was easy and free from error. The bulk of mankind are school-boys through life."

Feeling like a frustrated student was what Jefferson hoped to change, and he didn't want to stop at getting rid of shillings and pence, either. He envisioned a country that harnessed the power of tenths in all its measures, and more.

American Measures

Jefferson's work on coinage was one of two historically notable reports he delivered in the spring of 1784. The other also dealt with divisions of a larger whole, but rather than of the Spanish dollar, it was of that part of the nation we now think of as the Midwest but at the time was called the Great Northwest.

Perhaps the most important issue considered by the Continental Congress during its Chesapeake Bay sojourn was how to handle the real estate windfall granted by the Treaty of Paris, which ceded from Great Britain an enormous expanse of land stretching from the Mississippi River to the Great Lakes. Individual states like Virginia had laid claims to this property but were being pushed to relinquish them so the land could be turned into new states, cut up into parcels, and sold for the benefit of the federal government. In truth, the states had little choice.

The nation was tens of millions of dollars in debt, and without the power to raise taxes or tariffs, its only great potential for income was the sale of that land.

As chairman of the committee preparing the report on how to deal with the western territory, Jefferson set himself to the pleasurable game of filling in a blank map. He drew up over a dozen new states, making them as rectangular as lakes and such natural obstacles would allow; it was the geographical equivalent of decimal math. If you bundle a couple of his smallish states together, you can kind of see Michigan. A few of the names Jefferson came up with also remain, more or less, such as Illinoia. The rest—like Assenisipia—not so much.

For all the problems facing the new country, it was thrilling to create a nation from scratch. The different states had become laboratories of political thought, each constructing its own constitution, for which models were few. The idea of a pact between the government and the governed was the truly revolutionary part of the American Revolution; in the rest of the world, the authority of the state was derived from customs and crowns, not from natural law and the people. But not just in government was the idea of America being created. Noah Webster was maneuvering to invent an all-new American language.

In 1783, Webster put out the original edition of what would become the first publishing phenomenon in U.S. history. *The American Spelling Book* or "blue-backed speller" would go on to sell millions of copies and lead to the all-American creation of the spelling bee. What its author really hoped to do, however, was create a native language distinct from England's. This would happen naturally, Webster believed, in the way that Dutch, German, and English all sprang from the same source and over centuries had become mutually unintelligible; still, he aimed to speed the process along. In his future dictionaries, Webster would remove extraneous letters from words like "colour" and "publick." Partly he did this to rationalize spelling, but he was also creating a situation in which a person could "hear" a British accent while reading. For Webster, a common tongue was a binding agent for nations, and he wanted to make the American language more uniform—at least as far as pronunciation went—from New England to the Deep South.

Just as there would be an American way to write, so, too, would there be an American way to reckon. To Jefferson, Gouverneur Morris, and all others who enthusiastically embraced decimal money, it would be a step in leaving the Old World behind. They saw pieces of eight and the £sd divisions as the irrational corruptions of European despots and relics of the Middle Ages, which kept the common person at a disadvantage owing to the difficulty of calculation. Decimals would form a new republican math, which would—in Jefferson's vision—be taught to the innumerate populace by dimes and cents, thus paving the way for the next step: the decimation of all weights and measures.

The idea of applying decimal fractions to weights and measures was not new. Simon Stevin had suggested it in the appendix of *Disme*, and decimal systems had been created and debated by the savants of England, France, and Italy starting in the 1660s. By the 1780s, the use of decimals for even the clock and calendar had been advocated in the *Encyclopédie*, that precious collection of Enlightenment knowledge Jefferson had been so reluctant to part with.

As the base of his system, Jefferson chose the one universal measure then in use, what he called the "geographical mile." A mile was in origin a *mille passus*, one thousand paces of five feet each. By medieval days, the mile had lost its original meaning and occurred in wild variations across Europe, with some Scandinavian miles being twelve times the original, which itself was a little shorter than the present American one. In Italian cities, the mile hewed closest to its original length, due in part to the many "milestones" that Roman engineers had placed on roadsides. On the open sea, where one navigated by geographical coordinates, the mile had taken on a different significance, as an old sailor's rule of thumb held that a degree of latitude equaled sixty miles. (In English statute miles, it equals about sixty-nine.) Navigators the world over made use of this sea mile, with a *knot*, for example, measuring one nautical mile per hour.

The sea or nautical mile was the same as Jefferson's geographical mile, and what appealed to him beyond its wide use was the fact that it was almost exactly six thousand feet long. This gave Jefferson the opportunity to put the *mille* back in the mile, as he imagined setting it equal to one thousand paces of six geographical feet, which would hardly be different from the foot Americans were already using. As with the dollar, the most familiar measure would be essentially maintained.

Nothing in this was new—the idea of a universal system based on a mile of this sort was 125 years old—and that was the beauty of it. If all other countries built their systems from the same source, all of them would be compatible.

Jefferson spotted a place to use a geographical mile right away. In his plan for the western lands, Jefferson carved out states using degrees of latitude and longitude. "Michigania" ranged from the 45th north parallel to the 43rd, "Assenisipia" from the 43rd to the 41st, and so on. This meant that each was to be two degrees or 120 geographical miles tall. Jefferson wanted to then parcel the new states into lots of one square geographical mile, which he explained in his report contained "850 acres and 4-10ths," a confusing notion to people who knew a square mile to be 640 acres.

Unsurprisingly, Jefferson's geographical-mile-based parcels didn't make it into any of the ordinances dealing with the Northwest that Congress passed. This was in contrast to his dollar plan, which would largely be adopted.

When the first coins were struck, the tenth part of the dollar would be named, appropriately enough, the *disme*. The first *cents* appeared later, and the proposed *mills* never. There was also the *eagle*, a gold coin worth ten dollars that would long be as important as the silver dollar itself, if not more so. All was as Jefferson proposed, with one exception—there was no 20-cent double-disme coin. In its stead was the 25-cent piece, which was two bits of the old pieces of eight—a quarter—as well as nearly equal to a shilling. Jefferson may have been convinced that the people would ease rapidly into the new decimal currency, but not everyone was.

In 1784, however, the first American coins were still eight years in the future—a lifetime away in the tumultuous days of the late eighteenth century. By the time the Continental Congress had even adopted Jefferson's plan in 1785, the capital had moved twice since Annapolis (to Trenton and then New York), and Thomas Jefferson was no longer in America. He was instead in the place where so much of the tumult would be generated: Paris.

3/16

AMERICAN PARIS

I N THE summer of 1784, Paris was a city of sensations and savants. City residents were treated to the spectacle of the first woman flying, thanks to the balloons of the Montgolfier brothers, who the previous autumn had sent the first men aloft. In the salons of the aristocracy, Franz Anton Mesmer employed mass hypnotism in displays of what he termed Animal Magnetism. Was Mesmer a savant or a charlatan? Everyone had their own opinion.

The man considered the greatest savant of all was Paris's most famous and beloved resident; he also happened to be an American—Benjamin Franklin. In a day and place where scientific discoveries were a cause for celebrity worship, the man who was unsurpassed for his inquiries into electricity was also unsurpassed in exciting public interest. Franklin was a fad in and of himself. "No-one was more fashionable, more sought after in Paris than Doctor Franklin," said the artist Élisabeth Vigée Le-Brun. "The crowd chased after him in parks and public places; hats, canes and snuffboxes were designed in the Franklin style."

The celebrated Franklin was a warm welcoming face for his fellow American Thomas Jefferson when Jefferson arrived at summer's end, as was John Adams. The three of them had previously spent time together in Philadelphia in 1776 during the crafting of the Declaration of Independence, a document much admired by liberals abroad, especially in Paris. The French, however, didn't give credit for it to Jefferson, to whom the lion's share belonged, but to Franklin, as they did pretty much anything else they found to like about America.

The Philadelphian had been in the city for nearly a decade, having come over soon after the Declaration's signing to marshal French support for the Revolution. Whether he had done his job brilliantly or the French simply got involved because it was to their benefit (and satisfied the king's whim) was arguable; what was sure was that French intervention had been decisive in the conflict.

The then current perception of Franklin as a towering genius contrasts with the kindly, avuncular Franklin often cast as a folksy bit of comic relief in the story of our nation's founding. His buffoonish alter ego was partly his own creation, a literary device he employed to come off as not too serious or hectoring. Take for example Franklin's 1784 letter to the editor of the *Journal of Paris*, which begins with the author waking up at six in the morning to a noise:

> Your readers, who with me have never seen any signs of sunshine before noon, and seldom regard the astronomical part of the almanac, will be as much astonished as I was, when they hear of [the sun's] rising so early; and especially when I assure them, *that he gives light as soon as he rises*; I am convinced of this. I am certain of my fact. One cannot be more certain of any fact. I saw it with my own eyes.

(This tongue-in-cheek essay would be credited with inventing daylight saving time, as Franklin detailed the economy in candles and money that could be had by setting the clocks back to fool city dwellers into waking up with the sunlight.)

Parisians glorified Bonhomme Richard as a natural genius from the wilds of America, and Franklin knew how to play it up, dressing in simple rustic clothing that appealed to a French elite smitten with Rousseau. However he looked, Franklin was the toast of the town, with an ambitious social schedule that kept him a regular at the finest salons.

Jefferson hadn't seen Doctor Franklin in almost eight years and enjoyed being again in his company, as everyone did. Or nearly everyone. John Adams was not charmed by Franklin, despite—or because of—the long periods the two had spent negotiating various treaties together. Grudging respect aside, Adams had grown to detest Franklin. He found the old man a fraud, and he couldn't stand how the French ate up his fur-hat-wearing savant-of-the-wilderness act. It gnawed at Adams how seemingly all of Europe believed that Franklin had won the Revolution with a wave of his "electric wand." For his part, Franklin summed up his fellow founding father by saying, "He means well for his Country, is always an honest Man, often a Wise One, but sometimes and in some things, absolutely out of his Senses."

Adams and Jefferson, on the other hand, had been fast friends from the start, and they renewed their friendship in Paris, which continued after Adams got appointed minister to London. When Jefferson visited him they toured the English countryside together, where they stopped in the childhood home of Shakespeare and chipped off a splinter of what was supposed to have been the bard's chair, "according to the Custom."

Their families bonded, too, particularly Jefferson to Adams's seventeen-year-old son, the precocious John Quincy. Blessed with razor-sharp intelligence, "Little Johnny" had already served on a diplomatic mission to Russia. Jefferson loved being in the young man's company, and the feeling was mutual.

The true center of American culture in Paris did not revolve around either Franklin or Adams, however. That honor belonged to a man who wasn't American at all, at least not by birth.

It is hard to imagine that any Frenchman before or since has loved America so dearly as did the Marquis de Lafayette. Born to impeccably high status and extreme wealth, Lafayette was a toddler when he lost his father to a British cannonball during the Seven Years' War. Partly because of this, Lafayette came into his inheritance at an unusually young age. While still a teen, Lafayette followed an impulsive and romantic idea across the ocean, not only bent on avenging his father against the enemy English, but swept up in the idealism of the American Revolution. Thanks to the recommendation of Franklin, Lafayette got a commission and served at George Washington's side during the Valley Forge campaign, going on to become a hero of the Revolution and finding a replacement father in his commanding officer.

Lafayette kept America in his heart and on display at his extravagant Parisian home, with English-speaking children—a son named Georges Washington de Lafayette and a daughter named Virginie—as well as a young Native American servant decked out in full "Indian" regalia, who would perform dances for guests wearing nothing but strategically placed feathers.

Jefferson knew the marquis from the Frenchman's defense of Virginia, and upon his arrival, Lafayette put himself at the American's disposal. Most important were the contacts and influence the marquis could offer the new minister, whose assignment was to secure favorable trade treaties for the United States. For his own part, Jefferson wanted to break American reliance on trade with the hated British, and he saw France as his country's greatest potential partner. Soon Jefferson would be taking over Franklin's duties as ambassador, too. Nearly eighty, Poor Richard was feeling old and tired and wanted to go home. While Franklin was escorted to the port of Le Havre on the personal litter of Marie Antoinette, Jefferson learned that succeeding his friend at the court of Versailles was "an excellent school of humility." At his weekly audiences with the Crown, the minister from the minor new nation across the sea had to work to get noticed, which made Jefferson's relationship with Lafayette all the more vital.

The two grew close, with Lafayette coming to consider Jefferson among his closest friends. The pair kept regular company with two other liberal-minded French aristocrats, the Duc de La Rochefoucauld and the Marquis de Condorcet. Both had also been supporters of America's revolution, although their aid came through political writing rather than on the battlefield. In many ways the most intriguing friend Jefferson would make in Paris was Condorcet, the permanent secretary of the Royal Academy of Sciences and thus—quite literally—the leader of the savants.

The Academy, the Farm, and the Wall

In the day of the savant, the Academy of Sciences stood at the pinnacle of intellectual life. The Academy was a kind of aristocracy of science and an organization without parallel in the Anglo-Saxon world. London had its Royal Society, but it was royal only in name, being a private club of gentlemen. The Academy of Sciences was one of several French academies, all with the Crown's money and support behind them and housed at the palace of the Louvre. The first academy was established in the beginning in the seventeenth century and dealt largely with molding an official French language, succeeding by royal decree in doing what Noah Webster hoped to accomplish via cheap paperbacks. A few decades later, Louis XIV founded the Academy of Sciences, as much to gain military advantages as from wanting to advance natural philosophy.

Part of the Academy's mission was to serve as a gatekeeper to science. When Anton Mesmer wanted approval for what he believed to be epic discoveries, a commission stocked with Academy members convened to judge their validity. Cochaired by Benjamin Franklin, the commission found there to be none, to the dismay of both Mesmer and his elite clientele.

The 1780s were a time of one of the Academy of Sciences' great flowerings. A few months younger than Jefferson and still in his early forties, Condorcet had already spent a decade at its helm. The marquis owed his post as secretary in part to his work as a groundbreaking mathematician, but he was as much a political animal as anything. For Condorcet—as for Jefferson—science and politics didn't stand apart. Condorcet fervently believed that science and human rights marched inexorably together; their advance was progress, and from progress all happiness would follow.

The optimistic philosophy for which history remembers Condorcet stands at odds with his frenetic temperament. His passions emerged in radical political thought; he believed in equal rights for women and blacks and in giving the vote to everyone, not just propertied males. It was impossible for Jefferson to square his own beliefs with all of Condorcet's, including the hostility to organized religion bordering on atheism that the marquis—along with many other French savants—wore nearly as a badge of honor.

Condorcet had been working for the government since having a decade earlier been appointed head of the mint, the classic savant–statesman job going back to Isaac Newton. Significantly, one of his earliest tasks was creating a scientifically based system of weights and measures to be used across the nation.

The assignment was handed to Condorcet by the same man who brought him into government, Jacques Turgot. In 1774, Turgot had been appointed Minister of Finance, the most powerful office in the country. Turgot was associated with the Physiocrats, the classical liberals who formed the first school of economics. Closely tied to the term *laissez-faire*, the Physiocrats believed in free trade as a panacea for an economy's and even society's ills. In France, this meant dismantling the tariffs that were everywhere a part of French life. For his part, Turgot wished to create a more rationally administered nation. Hiring Condorcet and creating a new system of weights and measures were important steps in that direction, as was his recruitment of another member of the Academy, Antoine Lavoisier.

In the years Jefferson was in Paris, Lavoisier was at the peak of his powers, doing for chemistry what Galileo had done for astronomy; instead of a telescope, however, Lavoisier's transformative tool was the scale. Wealthy, haughty, and tireless, Lavoisier inspired jealousy and little affection, even among those who admired his achievements. Lavoisier did nothing to help his popularity by the particular pride he took in exposing scientific charlatans such as Mesmer, whose commission he cochaired with Franklin.

Of his own accomplishments, Lavoisier was sure. He even boasted that without him, American independence would never have happened. Working under Turgot in the mid-1770s, Lavoisier overhauled France's production of gunpowder, a material the French were so chronically short of that it was blamed for their humiliating defeat by Britain in the Seven Years' War. Lavoisier's methods—both chemical and organizational—proved so successful that France began running a surplus, which turned into a lifeline for American freedom fighters.

What France did not have a surplus of was cash. Turgot was against adding to the country's already deep debt and thus opposed also to French support of the colonial rebels. This advice was not to the liking of a young Louis XVI eager to stick a thumb in the eye of the British. Like Lafayette, he wanted to avenge the Seven Years' War, a fact that more than anything else—Franklin's charms included—led to his country's participation in the American conflict.

Turgot was soon sacked. Rather than launching a liberal revolution from within, he was just the first through the turnstile in what would be a long line of finance ministers for Louis. Although his reform of weights and measures was scrapped, Condorcet would stay on at the Mint, just as Lavoisier continued to lead the Gunpowder Administration. But Lavoisier's principal job—and the source of his copious wealth—was being one of a select group of shareholders in the most profitable concern in France, the General Farm. In a day when governments leased out monopolies and franchises (or "farms"), the General Farm had the best one of all—collecting taxes.

These were not taxes on income or property. At the time, governments relied on tariffs to raise revenue. In France, these were internal and levied unevenly in differing regions and cities, putting a chill on commerce. Even more hateful were the taxes the General Farm collected on necessities such as salt, a burden that fell disproportionately on the poor.

The Farmers' profit depended upon their efficiency, and for their efficiency they were despised. With police powers and well over twenty thousand agents, the Farm functioned as a state within a state, and an unwelcome one. The popular nickname of the Farmers—the *sangsues*, or bloodsuckers—gives a fair indication of what the people thought of them.

Far from being a passive member, Lavoisier put his brilliant mind and rigorous methods to the problems of the Farm, including his 1779 proposal for what would become its most massive project: the construction of a wall encircling Paris. Ten feet high, eighteen miles in circumference, the wall of the Farmers was approved and budgeted at a massive sum, to be paid for at the public's expense. Adding to its hideous cost was its elegant design, created by the neoclassical architect Ledoux, who imagined its every tollbooth as a grand entryway into Paris, although the more prosaic aim of the wall was to stop the smuggling of goods into the city and enforce the Farm's tariffs. The reason the crown agreed to such a thing was its crippling debt, the interest on which was eating up half the budget. This situation caused the government to lean ever more heavily on the Farm, which extended the crown cash in anticipation of tax revenue.

The Farmers Wall—and the Farmers themselves—were anathema to the liberal Physiocrats and like-minded souls such as Jefferson. To the fuming residents of Paris, the wall conjured up a prison, and everything that had gone wrong with France.

In the end, France's support of the American Revolution had indeed proven financially ruinous. Going to and from his regular visits to Versailles, Jefferson passed through the stylish tollbooths of the Farmers Wall, and it bothered him no end that his country's independence had at least partly been responsible for such an abomination.

Ever since Jefferson had taken over from Franklin, debate over the debt and how to pay for it had been intensifying in Versailles, and in 1786 the latest finance minister came to the king to tell him that he—and thus the state—was going broke. More shuffling of ministers and back-and-forth struggles ensued, with crisis coming two years later, when money problems were exacerbated by a disastrous harvest and skyrocketing bread prices fanned the flames of unrest. Add in the heat of August—the month that the Crown suspended payments on its debts—and the situation turned incendiary.

More than just broke, France was breaking down, and Louis took a drastic step he had been avoiding for years. He asked his subjects what they thought he should do.

One King, One Weight, One Measure

In 1789, most aspects of French life were governed by the *coutumes*. This term is usually translated simply as "customs," but not adequately, for the *coutumes* regulated everything from feudal duties and measures to dowries and capital punishment as they had evolved through practices and traditions that stretched back before Charlemagne to the ancient Franks.

As for the French people themselves, they were split into three "estates." These were the classic feudal divisions of the realm and still held in many other parts of Europe, although Britain had long since shorn itself of the clerical estate and counted only lords and commons. In France, the top of the hierarchy remained the Catholic clergy, while the Third Estate was the bottom one, comprising everyone from the humble peasant to the wealthy industrialist, and included twenty-four out of every twenty-five Frenchmen.

In times of crisis, a king could call for an Estates-General, a meeting of representatives from the three estates. A primitive form of parliament, this consultation with the different French castes was held at the king's sufferance, and its opinions were nonbinding. Even so, kings didn't care for them and their whiff of democracy, which is why one hadn't been called for 175 years. This made the Estates-General slated to open in May 1789 an epochal event, the lead-up to which saw the election of representatives and the assembling of the *cahiers de doléances*, or "notebooks of grievances," which were a chance for the people to air their beefs.

At the same time this was happening, the United States was in the process of seating the first Congress under its new Constitution. The Constitution was about as different a foundation of government from the *coutumes* as could be found, a single document that created a "supreme law of the land" that applied equally to everyone. The *coutumes*, on the other hand, changed everywhere you went.

"Is it not an absurd and terrible thing that what is true in one village is false in another?" Voltaire wrote. "What kind of barbarism is it that citizens must live under different laws? When you travel in this kingdom you change legal systems as often as you change horses."

Whereas each state entered the American union on an equal footing, in France every patch of land, from free cities to principalities, had entered the realm on particular terms enshrined by custom. The deals that nobles, cities, and regions had cut for themselves were called *privileges*. Wrapped up in these privileges were France's measures, the state of which boggled the mind of English-speaking visitors.

Great Britain had a single set of legal standards, a blessing that had been extended to its American colonies. For as bad as the situation in coinage was, weights and measures were so relatively uniform in the United States that a Frenchman might ask Thomas Jefferson if he was crazy to want to mess with them.

By some counts, there were 250,000 different measures in France. To see how that could be possible, take as an example the six-foot long *toise*, the principal standard of the French. A given town might have its own local *toise* but use the king's *toise* for state-related business and the *toise* of the regional capital for still other matters, as well as other sets of measures in the surrounding countryside, which varied according to the individual lords. Traveling merchants had to live by thick conversion manuals they lugged everywhere they went.

While it seems insane, realize that this was a world in which dialects changed from town to town, or between neighborhoods of the same city. What's more, these weren't measures as we think of them, but more akin to today's byzantine overlay of federal, state, and local taxes. In fact, they were deeply intertwined with taxes, in the form of tariffs, rents, and fees, which is why they were such a major source of fury among rural Third Estaters, who weren't concerned with the diversity of measures so much as their abuse.

Peasants paid their rent not in cash but in kind, a way of life that can be traced back to the fall of Rome and the collapse of coinage. Wheat farmers owed their lord a portion of their grain harvest, calculated by the *boisseau*, or bushel. This *boisseau* wasn't made up of so many quarts and so many ounces, it was one particular bushel and nothing else, kept by the lord. The thing was, these *boisseaux* seemed to grow suspiciously over time, along with the lord's take.

This made peasants rage. They lived and died by these bushels, and for them the true measures of the past were emblematic of a day when all was fair and just in the world. Realize that their complaint was different from that of the old man on the corner who complains about the price of milk compared to when he was a kid. Inflation was a concept that wasn't meant to exist in the Europe of three estates; prices, like the size of a bushel, were never supposed to change, not in one life, not in a hundred. The immutability of wages, prices, and fees was part of the order of the things, which the measures set by *coutumes* were supposed to help guarantee. What many peasants wanted was for the king to step in and take the bushels out of the lords' hands and replace them with his own measure, the *boisseau* of Paris.

The king's measures were part of the *coutume* of Paris, which served not just the city but a wide area of northern France. Not only were there three estates, there were two Frances. One was the medieval kingdom that Voltaire deplored, the other a modern nation. This modern France sat roughly in the Paris basin, which stretched up toward the Low Countries and contained factories and highly productive large-scale farms. This was a land where the people largely spoke one language and lived under one set of laws and measures.

For those who thought of Paris *as* France (a group comprised of Parisians and foreigners), the France of the vast interior came as a shock. It was poor, rural, and filled with people who spoke mutually unintelligible dialects; by one estimate, 90 percent of Frenchmen couldn't speak "French." This astonishing cultural diversity was seen by liberal reformers as a sickness, the cure for which was uniformity, that everything and everybody be made the same.

The fastest route to a uniform nation was through the king and his *coutume*, that is, the universal application of his laws and standards. (That this was also the *coutume* of Paris, where liberal reformers like Condorcet lived, didn't hurt either.) "One king, one nation, one law" was a political slogan with wide appeal.

Another was "One king, one law, one weight, one measure." This was not a call for something new but rather a desire to return France to the sort of national perfection that existed under Charlemagne, who in 789—exactly one thousand years before—had issued the first of his decrees unifying measurement in the land. (That such perfection was surely a myth mattered not at all.)

Of course, the reform of measures on a countrywide scale had been tried, and recently. Condorcet's 1770s plan for weights and measures had failed precisely because it asked to be the sole system in the land, an aim considered hopeless and illegal, given the primacy of the *coutumes* and the attitude of those who benefited from them. This was not just lords, but cities and regions who saw in their privileges protection and independence from a greedy, centralizing king at far remove in Paris.

The situation was indeed hopeless so long as there remained the *coutumes*, which kept the nation enmeshed in one giant interlocking system of differences, from the three estates to a quarter million unique measures. For it to change, the entire structure of France had to collapse. And in a place where the edicts of Charlemagne remained relevant ten centuries later, what were the chances of that?

A Conspiracy of Well-Intentioned Men

Contrary to most popular depictions, the French Revolution was not started by pike-wielding peasants starved for bread and eager to behead aristocrats. Grain riots enflamed situations, but they didn't topple kings, who had been dealing with such crises forever. Rather, it was the aristocrats themselves who did it. Not the lords who stayed up at night with barrelwrights to fraudulently enlarge their *boisseaux* and screw peasants out of a few more kernels of wheat. Those were members of the poor nobility who hadn't been benefiting from the galloping progress of the Paris region.

The aristocrats pushing for change were the ones who had been doing well in the modern world and were only too happy to give up their feudal privileges; after all, they had been evicting peasants from their lands for decades in order to practice more efficient agriculture or start factories. Others, like Lavoisier, had bought their titles and so never had a stake in the old *coutumes*.

These elites began to form themselves into societies. There were the Amis des Noirs, the "Friends of Blacks," a society of abolitionists enthusiastically supported by Condorcet and Lafayette, who tried to get their friend Thomas Jefferson to join. (And failed, for obvious reasons.) More in Jefferson's comfort zone was another club composed of much the same membership, the Society of Thirty, which got together in the period leading up to the gathering of the Estates in the hopes of influencing its outcome. Jefferson was a regular guest of the group, who were called the Fayettistes or the Americans but styled themselves the "conspiracy of well-intentioned men."

Many of the well-intentioned men were voted to the Estates-General, including Lafayette, who was poised to play a central role, and La Rochefoucauld. Condorcet failed to get elected, being too radical for the bulk of his fellow nobles. Turned away for a different reason was Lavoisier.

At its start, 1789 looked to be Lavoisier's year. He published his seminal work, *Elements of Chemistry*, which provided the first description of the conservation of matter, and he and his wife sat for an outsize portrait done by David, a masterpiece to be hung in the coming Salon. Lavoisier's electability, however, was ruined by his being one of the *sangsues* of the General Farm, and for devising their hated wall.

The dearest desire of reformist nobles was that the Estates-General would lead to France's becoming a modern, unified nation, in which the people would have inalienable natural rights and not be subject to the capricious nature of varying *coutumes*. To achieve this, they would need a legal compact.

In Lafayette's ersatz-American home, there hung on the wall a frame, one half of which displayed a copy of the Declaration of Independence, the other half of which lay empty. Here was where a declaration of French rights would go, and Lafayette hoped to be the man to write it. Two days before the opening of the Estates, he had dinner with a couple of Americans who had experience in composing such documents. One was Jefferson, the other a man fresh to the Paris scene but who had known Lafayette at Valley Forge—and clashed with Jefferson over coinage—the peg-legged rake Gouverneur Morris.

Morris had come to France to look after the business interests of his old partner Robert Morris, which hardly endeared Gouverneur to Jefferson, especially considering he was looking to restore a contract between Morris and the General Farm over a tobacco monopoly that Jefferson had nullified. Gouverneur, however, had proved himself more than a mere merchant at the Constitutional Convention in Philadelphia, where he skillfully drafted the final version of the Constitution and composed the preamble that begins "We the people."

Wasting no time, Morris got himself enmeshed in the Paris social scene in ways Jefferson never had, soon finding himself in a love triangle with a certain Madame de Flahaut and her longtime lover, the Bishop d'Autun. D'Autun—later known as Talleyrand—remains one of the most serpentlike creatures in history. But at the time, he was a liberal member of the Society of Thirty and had been elected to the

Estates-General, where he would play a pivotal role. As a cleric, he served as a member of the First Estate, despite probably believing in God about as much as Condorcet did.

These men were the veritable cream of the late Enlightenment, and most of them—Jefferson, Lafayette, Condorcet, La Rochefoucauld, Morris, and Talleyrand—attended the May 5, 1789, opening of the Estates-General, which convened with great pomp and circumstance in Versailles. At its sessions, the question of what the country should be doing quickly gave way to who should be giving the orders—the nobles and clergy, or the people?

Six weeks in, the people (or rather, those elected to speak for them) took matters into their own hands, declaring the Third Estate to be the National Assembly and sole legislative body of the country—essentially, they proclaimed the people to be the nation. After a somewhat farcical interlude that found the Assembly briefly occupying the king's indoor tennis court, the Assembly's ranks swelled, initially by priests crossing over from the First Estate. By month's end, the king bowed to the popular will and ordered that all estates be merged into the Assembly.

As the weather turned hot, the bread crisis worsened and events intensified outside Versailles.

On a feverish midsummer night, mobs in Paris turned on the most hated symbol of the day. Rioters took hammers, clubs, and picks to some of the finest new examples of architecture in Paris—the tollbooths of the Farmers Wall. Tax records put to the torch brightened the night sky.

The citywide violence lasted for days and culminated with the crowd focusing its rage on a single building, the Bastille, which rioters stormed and captured. Soon, heads on pikes were being paraded through the streets. While men like Morris found sleep impossible, Jefferson believed it all overblown, claiming to slumber in his home as quietly as ever. "The cutting off heads is become so much à la mode," he quipped, "that one is apt to feel of a morning whether their own is on their shoulders."

Emboldened by events in Paris, a peasant revolt gathered steam in the countryside, where houses of the nobility and clergy had been getting ransacked, fueling what became known as the Great Fear. In response, the National Assembly passed laws on August 4 that ended all feudal privileges—a shocking outcome. When the king consented to the laws, he was hailed by the Assembly as the "Restorer of French liberty," and the nation seemed on its way to a British-style constitutional monarchy.

The final weeks of August were to be monumental ones for human rights. In the United States, the Bill of Rights was passed by Congress on Monday the twenty-fourth; that Wednesday, the National Assembly approved the Declaration of the Rights of Man and Citizen.

The first attempt at the Declaration had been given to the Assembly by Lafayette shortly before the fall of the Bastille, in whose aftermath the general became one of the most powerful people in France. He was appointed commander of the National Guard, a newly formed organization charged with returning order to Paris, the most important commission going. The radical nature of the National Guard was that it took arms and control away from the king's men (often foreign mercenaries such as the Swiss Guard) and placed police power directly in the hands of the people. For his men, Lafayette designed the tricolor cockade that would become the symbol of France and of republican government itself. At the birth of what would be a revolution, Lafayette—the hero of two worlds—was the man of the hour.

For Jefferson, it was thrilling to experience a revolution unfolding before him for a second time. He thought of the American and French revolutions as joined, no matter that the French one would have to be different. A map of France afforded no blank spaces to freely carve up, and the appropriation of church property orchestrated by Talleyrand was the sort of event that could have no exact parallel in America, even if the aim was the same as for the western lands—to sell off newly acquired property to pay down the national debt.

In September, Jefferson hosted a dinner attended by his closest French friends—Condorcet, Lafayette, and La Rochefoucauld—plus Gouverneur Morris, who commented on what a fine stock of wine the minister kept. Even though the guests didn't know it, the gathering was a farewell. Jefferson would be leaving shortly to go on a planned six-month trip home to Monticello. But when he landed in Virginia's port of Norfolk, the papers had news he wasn't expecting, and Jefferson would never return.

$4/16$

or One Quarter

METRIC SYSTEMS

W HEN THOMAS Jefferson arrived home in late 1789, it was to a country that was different from the one he had left. The United States had given itself a mulligan. By replacing the Articles of Confederation with the Constitution, America had become less a league of separate states and more a unified nation, with a federal government possessing real power and a strong executive to boot. It would also be getting a new home. After one more session in New York, the capital would move to Philadelphia, where it would await construction of the Federal City.

Upon landing, Jefferson learned that he had been selected by President Washington to be the nation's first secretary of state. It was one of just three secretaryships and would give Jefferson a portfolio that included all the nation's foreign interests as well as several domestic ones. Although it was the most important appointment of his life thus far, Jefferson was reluctant to accept. Once home in Monticello, he hated leaving, especially to go to New York, which he considered "a Cloacina"—a sewer—"of all the depravities of human nature."

But there was no denying Washington. Jefferson departed Monticello and, after a journey slowed to a crawl by an unusual foot and a half of March snow, arrived in New York, where he became schooled in the difficulties of the local real estate market. After some effort, he rented a

less-than-agreeable house on Maiden Lane that he hoped to occupy by
May.

Shortly before, Washington had delivered the first-ever speech of a
sitting president to Congress, later called the State of the Union address.
Midspeech, he stated that "Uniformity in the Currency, Weights, and
Measures of the United States is an object of great importance, and will,
I am persuaded, be duly attended to." The currency part would be
handled by the treasury secretary, Alexander Hamilton, whose report on
the mint would follow up on Jefferson's still-not-acted-upon dollar plan,
while the drafting of a report on weights and measures was assigned by
Congress to the secretary of state. Jefferson was picking up right where
he had left off, on a subject he had never stopped thinking about.

Half a dozen years earlier, Jefferson's ideas had focused on the degree-
based geographical mile. The potential for its becoming universal
remained its greatest plus. A geographical mile would immediately fit in
with all the maps and globes that used degrees, minutes, and seconds to
divide the world. As the dollar familiarized people with decimals, such a
mile would educate the citizenry. Writing to Jefferson from Paris on the
topic of a U.S. system of measures, Gouverneur Morris suggested that
the nation's roads should have milestones marking geographical degrees.
People could begin to understand themselves not as moving from point
A to point B, but traveling across the curved surface of the earth.

Adopting a system in which a mile was based on geographical degrees
would instantly put America in sync with any other country that chose
to do the same. Denmark had already been on such a system for nearly a
century; its *fod* was theoretically identical to Jefferson's geographical
foot—both were 1/6000th of one minute of arc. There was one catch,
however—two systems being compatible depended on their agreeing on
the length of a degree, which is to say, the size of the earth.

The circumference of the earth had been measured with relative accu-
racy for more than a hundred years. The math for figuring it was basic.
All you needed to do was measure one degree of a meridian of

longitude—which is to say, a line running due north–south—and then multiply that length by 360.

The devil was in the details, of course. For one thing, it's damn hard to physically measure a full degree (about 69 American miles) while factoring for things like the curvature of the earth and elevation, not to mention all the obstacles that might get in the way, like mountains. Just being able to precisely locate the start and end points of a degree was a challenging astronomical feat.

Ballpark figures were easy. The first relatively modern earth measurement was made by a physician using a makeshift carriage odometer on the Paris-to-Amiens road, which ran in a fairly straight north–south line. All in all, his calculation was pretty good. Using the same basic meridian as the doctor's, an astronomer-priest named Jean Picard performed a survey in the 1660s that provided a more scientific measure of the earth's circumference.

But with many such surveys having been made and more to come, which should a nation pick? Would countries prefer measures based on their own meridians? Once measures were chosen, what would happen when newer, better ones came along? Denmark couldn't be expected to keep changing its miles and feet, nor could someone like Jefferson be expected to want to base the geographical mile on some outdated seventeenth-century measure of the earth.

These issues brought into question the suitability of basing a system of measures on a degree-based mile; the earth itself was too challenging an object. A natural standard was the dream, but what was needed was something more easily testable. Such a standard, fortunately, was believed to exist. It was called the *seconds pendulum*.

The Planet and the Pendulum

In a day when grandfather clocks are seen as kitschy relics, it is hard to conceive of what the pendulum clock meant when it burst onto the scene in the 1650s. Sundials then remained the most trusted time instruments for townsfolk (so long as it wasn't cloudy), while the hour-glass was vital for navigators. Clocks, on the other hand, stood on the bleeding edge of technology, the best of them used as high-precision tools for charting stars, which is why the hour was divided into the astronomer's minutes and seconds. Most mechanical timepieces, however, were still not accurate enough to justify their being outfitted with anything more than an hour hand, losing as they did between fifteen minutes and half an hour a day.

The first pendulum clocks, by comparison, lost a mere ten to fifteen seconds per day, up to a hundredfold improvement. Instantly, all other clocks became obsolete, and owners had the guts ripped out of old cases to be retrofitted with works regulated by a pendulum. Clocks now not only warranted minute and second hands but were driven by the second itself, which is how long it took the pendulum to beat one way.

The key attribute of the seconds pendulum is its length, which is a little more than 39 inches and part of the reason a grandfather clock is so tall. That a pendulum's length is related to its period—the time it takes to swing back and forth—had already been known for decades. It was believed to be a natural constant and so easily testable that it was quickly seen as a candidate for a standard of measure against which all others could be checked. After the first pendulum clock was created in 1656, the idea became more popular, as the pendulum became iconic in and of itself.

As did the pendulum clock's creator. Christiaan Huygens was a handsome Dutch savant who looked something like a 1970s rock god in the long blond curls of his fashionably long wig. After studying at the University of Leiden, Huygens made a splash by discovering that what Galileo had called the "ears" of Saturn were in fact rings, and was still only in his midtwenties when he invented the pendulum clock. Much sought after outside the Netherlands, Huygens became an early member of London's Royal Society, an organization he was soon hired by Louis XIV to help create a French version of.

Huygens was an integral part of the Academy of Sciences during its phenomenal first period, when the Sun King assembled an international dream team of savants. In addition to Huygens, there was Giovanni Domenico Cassini, a formerly Bologna-based astronomer who would solve the problem of calculating longitude on land; the *wunderkind* Ole Rømer of Denmark, who was the first to show light had speed and to calculate it; and, from France, the earth measurer Jean Picard.

It was in this era that universal measures became a hot topic in the Republic of Letters, as the borderless intellectual sphere of Enlightenment correspondence is known. To share research, savants had to resort to swapping paper rulers back and forth in the mail to show the length of the particular *toise* or yard they were using, a patently ridiculous situation. It was for this reason—making results mutually intelligible—that the idea of the pendulum as a natural standard became popular.

Instead of sending rulers, savants could swap results as checked against a seconds pendulum. That is, they could define their yard or *toise* in terms of the length of a seconds pendulum, a tool that most kept at hand. Some thought the pendulum itself should be made the international standard; Huygens suggested that a third of it be called a *pied horaire*, a "time foot." Others pressed for using new and seemingly accurate measures of the circumference of the earth as the standard, an idea that inspired the Danish *fod* and Jefferson's early efforts.

Both the circumference of the earth and the seconds pendulum, however, proved to be problematic. In 1672, a member of the Academy running a series of astronomical experiments near the equator discovered that his pendulum clocks were running slow. To make them keep time correctly, he had to shorten the length of their pendulums by nearly an eighth of an inch, a discovery with massive ramifications for the history of science. The mystery bewildered most everyone but Isaac Newton; for him, the slow pendulums served as evidence for his controversial theory of gravity. Weaker gravity caused pendulums to lose speed; the reason *why* gravity was weaker at the equator than at Paris (at 45°N latitude) was that the earth bulged in the middle. Up until then, it was believed that the earth was a perfect sphere, meaning that every segment of every meridian followed the same curve; if that wasn't true, all estimates of the earth's circumference were wrong.

But if the earth wasn't a sphere, what shape was it? The debate, one of the most intense in scientific history, was still not entirely settled in 1790. As the question affected any calculation of the size of the planet, it made basing a measurement system on a meridian survey problematic. But the fact that one's latitude affected the performance of a seconds pendulum was a knock against the *pied horaire*. The issue thus complicated, savants had argued back and forth for more than a century over which made the better natural standard, the earth or the pendulum.

While Jefferson was in Paris, he found his opinion swayed away from the degree-based geographical mile. His friend Condorcet was a major advocate of the seconds pendulum, which he had used as the standard for the measurement system he had proposed in the mid-1770s. The seconds pendulum remained the most easily testable natural standard—differences in latitude could be taken into account without much difficulty—and it also was more neutral than any earth-based measure, which depended on surveys conducted in a given country. For these reasons, it had emerged as the international consensus, and so it would be around the pendulum that Thomas Jefferson would build a new system of weights and measures for America.

Measurement Spring

The first weeks Jefferson spent in New York were filled with a busy schedule of social engagements, some of which were even to his liking, such as visiting his old friend John Adams, the nation's first vice president. As the calendar turned to May, however, things turned dark for the secretary of state. He was struck down by an attack of the migraine headaches that plagued him at irregular intervals throughout his life. The house he had rented was not ready on time, so it was in the confines of a boarding-house that he suffered through blinding headaches and struggled to write the weights and measures report he was so anxious to present.

The plan he was putting together stuck largely to long-established principles, but did, however, offer one innovation—a better kind of pendulum. Information on this better pendulum had come via a close friend who shared Jefferson's mania for measure, James Madison. In France, Jefferson had procured a pedometer for Madison, which Jefferson himself had made liberal use of before passing on. A member of the Congress that had assigned Jefferson the job of fixing the nation's system of measurement, Madison agreed with many of his friend's opinions on the matter, and he believed the multiplicity of measures to be second only to the different languages as a divisive factor in the world.

As for the seconds pendulum, Madison gave Jefferson a pamphlet from a Philadelphia clockmaker named Robert Leslie that outlined how using a solid rod rather than a swinging bob would make pendulums less susceptible to gravity. The horologist believed that this seconds rod should be the universal standard of measures, and Jefferson agreed. What helped clinch it for the Virginian was the length of this rod, which was very close to five feet.

Whereas the English foot would grow more than an inch if set to a third of a seconds pendulum, it would shrink by only a quarter inch if based on a fifth of Leslie's rod. As with the Spanish dollar, Jefferson believed it was vital to center his system on the most familiar measure and then decimalize its divisions and multiples.

But what would replace the pound and pint? Any rational system had to be integrated, Enlightenment thinking went, meaning that all measures needed to be derived from a single source. The best source was considered the standard of length, which was what made it so crucial.

As for how a linear measure can be used to create all other measures: Length squared defines area, and length cubed defines capacity. Weight, in turn, is derived from capacity. *A pint's a pound the world around* comes from the idea that a pint of water weighs a pound. This relationship had ancient roots, with the Romans having based the original pound on the weight of water.

Jefferson decided to set the bushel equal to one cubic foot and use it as his standard of capacity; it was about 25 percent smaller than the standard Winchester bushel, but with lots of conflicting capacity measures running around, the match was not critical. With the pound, Jefferson had a thornier problem. No decimal fraction of the weight of the new bushel filled with water came anywhere close to matching the common pound. A one-thousandth part of a cubic foot of water, however, weighed almost exactly an ounce. What was even better, that ounce was very nearly the same weight as the silver dollar, a double discovery that proved to be the clincher.

Jefferson had his plan. One fifth the pendulum rod would equal the foot, one cubic foot would make the new bushel, a bushel of water would weigh a thousand ounces, and an ounce of silver would be the dollar. All other standards would be slotted in decimally, taking the names of close customary equivalents. This resulted in some awkward rejiggerings, such as the ten-thousand-foot mile being nearly twice its old length and the ten-ounce pound weighing half of its former self, but there was no way everything could work out perfectly. Jefferson had done about as good as could be hoped.

But would Congress agree? Fearing they would never go for such a radical scheme, Jefferson gave legislators an alternate plan whose principal aim was just making American weights and measures uniform. This was uniformity in the sense of a pound of feathers and a pound of gold weighing the same, and a pint taking up the same cubic space no matter whether it contained water, milk, or strawberries.

With weight, Jefferson essentially combined two existing systems, using the specialized troy pound (gold) for the low end and the general avoirdupois pound (feathers) for the high end, while just discarding the ancient apothecary weights altogether. Few other than pharmacists would notice the difference.

THOMAS JEFFERSON'S DECIMAL SYSTEM
OF WEIGHTS AND MEASURES

Unit	Value	Equivalency in English Customary Measures
Length	*Decimal feet*	
Mile	10,000	1.85 miles
Furlong	1000	1.48 furlongs★
Rood	100	97.87 ft
Decad	10	9.79 ft
Foot	1	.98 ft (11.74 in)
Inch	.1	1.17 in
Line	.01	.12 in
Point	.001	.01 in

★English customary furlong = 1/8th mile or 660 ft

Volume	Decimal bushels	
Last	100	93.76 cu ft
Quarter	10	9.38 cu ft
Bushel	1	.94 cu ft or .75 bushel
Pottle	.1	.7 gallon (wet)
Demipint	.01	.56 pint (wet)
Metre	.001	.9 oz (wet)
Weight	*Decimal pounds*	
Hogshead	1000	586.02 lb
Kental	100	58.6 lbs
Stone	10	5.86 lb
Pound	1	.59 lb
Ounce	.1	.94 oz
Double-scruple	.01	41.02 grains
Carat	.001	4.1 grains
Demi-grain (or Minim)	.0001	.41 grains
Mite	.00001	.04 grains

His change for capacity measures was more drastic, and had to be. Even today the United States has two kinds of volume measures, one for liquids and another for dry stuff. In 1790, there was a separate gallon for beer and dairy, as well as sole-purpose barrels and bushels for items like pork and coal. What was more, individual measures varied from state to state, or even within states. To eliminate this mess, Jefferson proposed a single gallon that was halfway between the two principal existing ones. This standard gallon would be used for all substances, with other capacity measures slotted in at different multiples.

Jefferson finished his report toward the end of May but held it back to tweak it. In the meantime, things began to look up for the secretary of state. At the beginning of June, he finally got to move into the "mean" house he had rented and his migraines abated. After having given the country a scare with a serious case of pneumonia, President Washington was also on the mend, and the two Virginians celebrated their improved health with a three-day fishing trip off Sandy Hook, near the entrance to New York's harbor.

Once back in the city, Jefferson decided to send off his proposal for a bit of peer review, with one copy going off to the astronomer and instrument maker David Rittenhouse and another to Alexander Hamilton. Just two days after mailing them, however, he learned of a proposal presented to the French National Assembly that outlined a scientifically based system of weights, measures, and coinage not just for France, but for the world.

That France would be overhauling its measurement system came as no surprise—it was a practical necessity. With the abolition of feudal privileges the previous summer, the old system had been largely dismantled, with nothing designated to fill the void. The assumption by many was that the measures of Paris—the king's measures—would be implemented countrywide, as had indeed been formally proposed.

The man who put forth the more radical proposal was Talleyrand, who was then being vilified by Catholic sympathizers as the "lame devil,"

"Judas," or just plain "Satan." Not only had Talleyrand proposed the confiscation of church property and its sale, but the soon-to-be-excommunicated former bishop was playing a key role in setting up the civil clergy, which forced priests to break from the pope in order to swear loyalty to the French state.

To the Assembly, Talleyrand cast a universal measurement system as part of the larger and ongoing reorganization of France. Not only had feudal privileges been abolished, but so had the nobility itself; not only had church property been nationalized, but monastic vows had been prohibited. Even the map of France had been annihilated and remade.

Gone were such ancient names as Normandy, Brittany, Provence, and Burgundy, along with their traditional borders. In their place eighty-three departments of roughly uniform size were established, with names that evoked natural features, such as High Rhine, Low Rhine, High Pyrenees, Low Pyrenees, Eastern Pyrenees, and Mont Blanc. Even if the geometry wasn't quite so relentless as Jefferson's latitudinally and longitudinally derived western states, they were still modeled along the same principles and scored another victory for uniformity over diversity.

Talleyrand hit hard the idea that the new France must not just be incrementally improved, but the very model of enlightened progress and a shining beacon for all mankind. Taking the king's measure would be the simplest course, he allowed, but not the best. Ultimately, those measures were arbitrary, whereas a scientific standard could be derived from a higher authority, nature itself. Such a standard would be neutral, whereas the Parisian measures of the king reeked of centralization and takeover. This new system would help unite not only France but the world. In the same way the Rights of Man were not just for the needs of Frenchmen in the late eighteenth century, but for all men, forever, so, too, should the system of measures be.

Talleyrand called for using the seconds pendulum as the standard, but he didn't push for decimal measures, seeming more inclined to stick to the duodecimal system that France already mostly used. Essentially, it was the same plan as Condorcet had cooked up in the 1770s, and the hand of the Academy secretary was likely behind Talleyrand's proposal. Talleyrand, after all, was a master of political manipulation, not a savant.

In order to make the new measures truly universal, Talleyrand desired British cooperation, proposing that the project be jointly undertaken by the French Academy and the British Royal Society. He saw this as leading not only to a cross-Channel system of weights and measures but also to a single market of free trade. In this way, two different kinds of borders could be erased at once.

For Jefferson, receipt of Talleyrand's proposal was fortuitous. A few days later, he received word of yet another proposal, this one made in the British Parliament and even more similar to his own, as it called for a seconds-pendulum-based system of decimal measures. The coincidence was inspiring.

In order to put his proposal more in line with the expected French one, Jefferson changed one detail of his plan, relocating his base calibration for the seconds standard from 38°N (the median latitude of America) to 45°N, the level of Paris and northern Maine. Jefferson didn't, however, abandon Leslie's rod for the seconds pendulum, surely hoping the American innovation would win over men like Condorcet and Talleyrand, both of whom he sent copies of his finished plan to.

In submitting his report on weights and measures to Congress on the fourth of July 1790, Jefferson took the unusual step of asking the Congress not to approve it. At least not yet, in "due regard to the proceedings of other nations engaged on the same subject." It was the wise move. What was the point of the United States' launching a system if their two greatest trading partners (and two most powerful nations on earth) came up with a different one destined to become the international standard? Besides, so much had gone so well in what was now a sister revolutionary state, and his confidence in the French was so high, that the secretary of state had little reason to believe that he and the rest of America would not like what was already being called the *meter*.

The Academy Decides

Of the various proposals submitted for the reformation of French weights and measures, the National Assembly chose Talleyrand's. It would not be up to Talleyrand to shape the new system, however, as such a task devolved naturally to the Royal Academy of Sciences. The first question the savants of the Academy needed to answer was a mathematical one: Should the new weights, measures, and coins of the nation be decimal, or derived from some other numerical base?

The committee put together by the Academy in the fall of 1790 to decide this issue was as great as any ever assembled in terms of mathematical achievement. Mathematics was the strength of the Academy at the time, with no fewer than six major historical figures in the discipline, half of whom served on the commission, including Condorcet.

But it was not only mathematicians at the table. Lavoisier was also involved, and it was no secret where his position on the issue lay. In *Elements of Chemistry*, Lavoisier admonished his fellow savants to use decimals when recording measurements and not break down results into "arbitrary" divisions, by which he meant ounces, scruples, drams, and so forth.

As hard as it is to imagine that Lavoisier and this committee of math demigods would have rejected decimals—the very engine of Hindu-Arabic numerals—support did exist for other systems, and long had. For good reason. Other than its being the number of fingers people possess, ten doesn't have much to recommend it as a number to base your counting system on. Its lack of divisors (particularly its inability to be split into thirds and quarters) is the reason why cultures everywhere tried just about everything *but* the number ten for their measures. In the West, tenths had been little used outside the church, with its *tithe* (from the

Old English for "tenth"), and the Roman army, where every tenth soldier of a mutinying legion would be killed, the origin of the word "decimate."

Base-12 being generally considered superior to base-10, some felt that rather than duodecimal measures being put in line with decimal numbers, numbers should become duodecimal. In a time of extreme change, what could be more radical than adding two new Arabic numerals and setting 10 equal to a dozen? Under such a system the year would have 10 months, each half of the day 10 hours, a *sou* (shilling) be made up of 10 *deniers* (pence), and a foot composed of 10 inches of 10 lines of 10 points.

But the days were not quite so radical yet, and although such a counting system was seriously proposed, adding two numerals to the Arabic ten was not a step academicians were willing to take. Nor were they willing to consider shaving off two numerals, as some advocated, seeing in base-8 the octal math of the people, let alone harebrained base-2 and base-11 schemes.

For the most part, the savants didn't need to be convinced that weights and measures should be decimalized. They believed in decimals—even coming to see the accident of the hand as a true natural standard. Across the ocean were men of similar belief.

Although no U.S. coin had yet been minted, the dollar was already being lauded. One Philadelphia pamphleteer extolled the joy of being able to multiply and divide the new money "by dots." Others would soon talk of the new republican math and how the "tree of liberty" was again putting forth a blossom in America. Decimals became about democracy; Gouverneur Morris talked of the importance of people understanding measures so they could participate in government, just as Condorcet believed decimals would ensure that "all citizens can be self-sufficient in all calculations related to their interests; without which they can be neither really equal in rights . . . nor really free."

With the fundamental decimal question decided, a second Academy committee met to pick what the new universal standard should be based on. In essence, they were to settle once and for all the hundred-plus-year-old debate between the seconds pendulum and the earth. To most, the pendulum seemed as obvious a choice as decimals, but it was the force with which decimals had been embraced by the first committee that caused the savants on the second one to veer from their expected course.

The issue was their conviction that *all* measures needed to be decimalized. That included time and geographical degrees, which would invalidate the seconds pendulum and degree-based mile as then calculated. The second was

$$\frac{1}{86,400} \left(\frac{1}{2 \times 12 \times 60 \times 60} \right)$$

of a mean solar day and the geographical mile

$$\frac{1}{21,600} \left(\frac{1}{360 \times 60} \right)$$

of the earth's circumference. Such supremely irrational fractions surely had to be eliminated, members of the committee believed, and the earth and day divided into tenths and hundredths.

That said, those divisions of time and the circle were already used across the Western world. If universality was the main goal, then to decimalize them would be to go backward, introducing diversity where none existed. What was more, to base the French system on the seconds pendulum or degree-based mile would boost the chances of the new system's being adopted outside France.

There was also the cost. Decimalizing astronomy and geography would mean scrapping all maps, star charts, and more. But how could members of the Academy—astronomers and geometers so many of them—ask any less of themselves than of the rest of the world?

That decided, the savants still had a choice to make. Would they use a pendulum calibrated to a decimal second, or a portion of the earth as divided by 400° of 100 minutes of 100 seconds?

They chose the earth. In part, this was because it seemed less controversial to introduce the concept of a 400° globe than a decimal second. Even better, the new division of the earth's circumference could be decimally divided to produce a unit of almost exactly three Parisian feet, or half the *toise*, the most important measure in both France and the Republic of Letters. (The Academy's savants were no less prone than Jefferson to such hometown rigging, even if they were ostensibly going for a universal system.)

Of course, the reasons why savants had soured on using the earth as a measurement standard hadn't gone away; its circumference was not exactly known and any calculation of it would change with advances in measuring tools. It just so happened, though, that one of the members of the committee, Jean-Charles Borda, had made a major advance in surveying equipment that promised the most accurate measures ever made, so much so the savants believed a definitive measure of the earth could be achieved. This appealed especially to Pierre-Simon Laplace, a physicist on the committee who sought to settle once and for all the true "figure of the Earth" (i.e., its shape).

In making their recommendations to the government, the Academy declared that the new unit would be the ten-millionth part of a quadrant of the earth's circumference, as determined by a new survey of a meridian. To decide which of the infinite number of meridians on the planet should be used, the savants said they had developed a set of criteria that only one could meet, which just so happened to be the meridian running through Paris that had already been surveyed multiple times.

To explain their dismissal of a time-based standard, the seconds pendulum was called "arbitrary." Despite that, the committee recommended that experiments be made under the supervision of the Academy to determine the seconds pendulum's precise length so that it could be used as a check on the future meter. In all, the committee recommended six separate scientific projects, the most important of which involved the execution of the meridian survey and an investigation into the temperature at which water reaches maximum density. The latter experiment would be overseen by Lavoisier and used to calculate the new standard of weight.

Taken together, the Academy was seeking with its recommendations not only the creation of a new system of scientific measurement but the advancement of science itself. It was a sign of the boldness of the times that such an ambitious plan could be approved by the Assembly, as indeed it was in March 1791.

Few outside the committee rejoiced.

A *Gâteau*

When Thomas Jefferson asked the Congress to table his weights and measures plan, he was putting his own country's business on hold because of his faith in the French Academy. It would have been hard not to feel that that faith had been abused, considering the plan they delivered.

How could the savants on the committee to select a standard be confronted with two universal options that had been debated for well over a century and choose *neither* of them? How was it that the year after Talleyrand recommended the seconds pendulum to the Assembly it was now being deemed "arbitrary"? How could it be any more arbitrary than choosing a forty-millionth segment of the earth's circumference as the basis for the meter?

In a letter to Condorcet, Jefferson reminded his friend why a degree-based mile had fallen out of favor in the first place—it couldn't be measured accurately. A survey that covered a tenth of a quadrant of a meridian, as had been proposed, still left the other nine tenths of it unmeasured and to be only guessed at, seeing as the figure of the Earth had yet to be decisively proven.

The secretary of state also took the secretary of the Academy to task over the lack of neutrality the committee had shown in choosing a

French meridian, which could be checked only within French territory. Though Jefferson didn't say it, the criteria they had come up with had obviously been tailored to suit the meridian they wanted to use; that there could be just a single meridian fit to be measured was absurd on the face of it.

The supposed need for a new survey was what provoked the most controversy in France. The king himself was baffled. This same meridian, after all, had been previously surveyed numerous times, and how could anyone be so sure that a new measure would be much more accurate? Besides, no matter how improved the latest tools, wouldn't there always be better ones to come?

It also didn't sit well that the committee included the toolmaker Borda, who stood to benefit the most from a new survey. Ulterior motives were detected everywhere, including the vested interest of Laplace in wanting to determine the shape of the earth.

And then there was the money.

The six ongoing research committees being created to establish the new measurement system would provide jobs to a dozen Academicians. The new weights and measures commission immediately received 100,000 livres, an amount three times greater than the annual funding of the entire Academy—and that was just a down payment. How could such an expenditure be justified when the debt crisis and bread riots that had brought France to revolution loomed as great as ever? It led some to think the measurement project was nothing more than an excuse to save Academy salaries. Funneling money into the pockets of an organization manned by aristocrats—as anyone in the elite was now being called, of noble blood or not—was unseemly at best, and dangerous at worst, especially considering what was going on in France during the summer of 1791.

The mood of the nation took a turn that June when a failed escape from Versailles by the king led to his becoming a de facto hostage of the National Assembly. He was all but forced to accept a constitutional monarchy, which wasn't enough for radical forces such as the Jacobin Club, whose members were calling for an end to the monarchy altogether. In this milieu, no voices were speaking louder than those belonging to Maximilien Robespierre and the scabrous journalist Jean-Paul Marat.

Marat took vitriol to soaring heights, tirelessly outing those he deemed Enemies of the People while styling himself the Friend of the People, the name of his newspaper. Marat hated the elite, but he held an especial loathing of the Academy in general and Lavoisier in particular. The organization had rejected (and Lavoisier dismantled) the scientific theories and experiments of Marat, who prior to the Revolution had been a physician. Far from being alone, a legion of frustrated inventors and would-be savants shared Marat's bitter resentment of the Academy's position as the gatekeeper of science.

Revenge came with the publication of Marat's forty-page pamphlet, *The Modern Charlatans, or Letters on the Academic Charlatanisms*. In it, Marat took direct swipes at the new measurement system, specifically the money being allotted to the Academy, which he called "a little *gâteau* to be shared among the brotherhood."

But beyond even money was the question of time. The scheduled scientific projects would take years to complete, while France needed a new measurement system right away.

To the savants who had worked up the plan, such complaints missed the point. They hadn't sought to address the mundane concerns of the day. Rather, their metric system was to be, as Condorcet put it, "For all time, for all peoples."

5/16

THE DECIMATION OF EVERYTHING

The best Picture I can give of the French nation is that of Cattle before a Thunder Storm. And as to the Government, every Member of it is engaged in the Defence of himself, or the Attack of his Neighbor.
—Gouverneur Morris to Thomas Jefferson, Paris, June 10, 1792

LIKE SO much else to do with the French Revolution, the guillotine emerged from a stew of rational ideals and grand symbolism. Proposed by Dr. Joseph-Ignace Guillotin, a physician who had served with Lavoisier and Franklin on the Mesmer commission, the idea was to create a humane death machine, if such a thing could ever be more than an oxymoron. To political minds of a certain bent, it represented another elevation of the average citizen, in that beheading had been the province of the elite—a noble death—while common hanging was a disgrace. Here uniformity and death, the two great levelers, collided.

When the guillotine was adopted on March 20, 1792, no one had any clue just how much use it would be getting. In the two and a half years that had passed since the Estates-General, the country had largely been in the hands of the self-styled well-intentioned men. Condorcet had done a turn as president of the revolutionary assembly, while Lafayette was enjoying as much power as anyone in the new France.

War changed things. From the start, the specter of battle loomed over the Revolution, the whole idea of which was anathema to the other crowned heads of the Continent, a generally more despotic crew than the

French sovereign. Conflict was stoked from abroad by nobles who had begun to flee France in the summer of 1789, with war finally arriving in April 1792 when France proactively invaded the Austrian Netherlands to attack the coalition forming against it. Things started badly, with the murder of a French general by his own troops and substantial losses in the field.

It was at this time that Gouverneur Morris became the American minister to France. In his first dispatch to his new boss, Secretary of State Jefferson, Morris sketched out a picture of scared government ministers turning on one another. In his second, penned a week later, Morris painted an even more dire portrait. "On the whole, Sir, we stand on a vast Volcano. We feel it tremble, and we hear it roar but how and when and where it will burst and who may be destroy'd by its Eruptions is beyond the Ken of mortal Foresight to discover."

Jefferson hadn't wanted Morris for his old job; in a list of three reasons why, the first was "his general character." But Washington had long been friends with Morris and appointed him over the secretary of state's objections. Jefferson's real problem was that he believed Morris to be a lover of monarchs and hater of republics, and he thought he would give the president a negative spin on the events taking place in France that Jefferson himself was so enamored of.

The volcano spat lava through early summer and erupted with the Brunswick Manifesto of July 25, 1792, a kind of final warning to France by the commander of the allied Prussian and Austrian forces congregated at the French border. The manifesto promised safety for the French people, so long as they left the royal family unharmed. Woe be to them, however, should any harm come to Louis and his clan, as the Duke of Brunswick promised a vengeance would be visited upon Paris that was to be both "exemplary and forever memorable."

The manifesto made the king's situation untenable. The emperor of France's principal enemy, Austria, was Louis's brother-in-law, while the king's own brothers were organizing counterrevolutionary forces to support the invading armies. These forces had as their principal goal the shredding of the Declaration of the Rights of Man and Citizen and the restoration to Louis of absolute power. Too many wondered if they had the enemy for a head of state, the king's protests to the contrary.

Fury at the royal family culminated in a heart-racing six-week period.

The Tuileries Palace was stormed by an armed mob under the direction of the increasingly radical Georges Danton, with hundreds of the Swiss Guard slaughtered and the king and queen taken into custody soon thereafter. A wave of mob violence by the *sansculottes* ("without breeches," the eighteenth-century equivalent of "blue collar") known as the September massacres followed, during which the National Convention formed, the latest successor to the legislative assemblies that had begun with the Estates-General. But this would be the first government without the king at its head. On September 22, the Convention abolished the monarchy and proclaimed the French Republic.

This establishment of the republic on the autumnal equinox of 1792 was the decisive moment of the Revolution—the point at which France separated from its thousand-plus-year past and became ruled by its people. It joined the United States as the second such nation, and men like Jefferson cheered.

A new kind of order it truly was. The faction later known as the Jacobins continued to rise, with its leaders Danton, Robespierre, and the ever more emboldened Marat finding common cause in those they opposed, a list that began with anyone who supported a constitutional monarchy. It was an ominous turn for aristocrats, no matter how liberal.

Already, well-intentioned men were fleeing.

Lafayette bolted in the days that followed the storming of the Tuileries, having been unable to duplicate his "father" Washington's success in forging a nation, a failing often blamed on what Jefferson termed his "canine appetite for fame." During the September massacres, La Rochefoucauld barely made it out of Paris before being taken by the mob and stabbed and stoned to death in front of his wife and mother. Talleyrand, in England trying to negotiate a peace (and being no more successful at it than he'd been in getting the British to join the meter project), decided it better not to return home, instincts that proved correct when a warrant was issued for his arrest toward the end of the year. The ex-bishop would eventually make his way to America, the place Lafayette most dearly hoped to end up. Instead, the Hero of Two Worlds was captured by the Austrians and imprisoned for over five years, despite the best efforts of Congress, Gouverneur Morris, and George Washington to spring him. Washington did manage to protect his godson and namesake, Georges Washington de Lafayette, the only member of Lafayette's family to escape. Young Georges would eventually make his way to Mount Vernon, where hung the key to the Bastille that his father had sent to the American president in more optimistic days.

Others should have been so lucky to get out of the country alive at all, starting with the king.

In December, "Citizen Louis Capet" was put on trial. Robespierre, a formerly unassuming lawyer from the provinces, was among those who believed it vital that the former monarch be put to death, stating, "Louis must die, so that the country may live." Condorcet—an opponent of capital punishment—thought Louis should be subjected to the worst job of the lowest citizen and confined to row in the galleys. Robespierre's prescription was more in the spirit of the times.

For American observers, the question of when things went wrong in the French Revolution depended on one's point of view. For some, like Gouverneur Morris, it was the earliest days, when he saw heads being paraded on pikes in the summer of 1789; for others, it was the January 1793 execution of Citizen Louis; and for those such as Jefferson, things had yet to go wrong, no matter what had become of his dearest French friends. Except, of course, when it came to measurement.

The Language of Revolution

One thing had become clear to Americans contemplating what sort of measures to adopt: there was no international system in the offing worth waiting around for. Not only did Thomas Jefferson want nothing to do with what the French Academy had cooked up, there was no British enthusiasm for it, either, nor any kind of decimal system. However unfortunate the opportunity missed, the unacceptable French plan did pave the way for Jefferson's report to be considered on its own merits, which a Senate committee did. In April 1792, the secretary of state's plan for a decimal system was unanimously recommended. Legislation putting it into effect, however, got bogged down that fall and winter, President Washington's regular agitations to get something done notwithstanding.

The lack of immediate action couldn't be too great a concern, or wholly surprising. After all, as dire as the situation in coinage had been, nearly a decade had passed between Jefferson's dollar proposal and the first coins being produced, an event that took place in the summer of 1792 with the opening of the Philadelphia Mint, the first federal building ever constructed in the United States.

In France, on the other hand, it was the situation with weights and measures that was most urgent, it being nearly three years since the old seigneurial measures had been abolished. Most critically, the lack of a national or even a legal system of measurement was disrupting food supplies. At the same time the U.S. secretary of state's plan was getting endorsed, the French interior minister, Jean-Marie Roland, was threatening to adopt the measures of Paris if the metric ones weren't soon in coming. All Roland cared about was having a single national system; which one it happened to be didn't much matter.

The savants on the weights and measures commission could have given the interior minister what he wanted, namely, the length of a provisional meter around which the system could be built. What worried Academicians such as Lavoisier was that a temporary meter based on a previous survey would be deemed good enough and the new meridian survey be abandoned, compromising their plans. But more than just its plans, the Academy needed to be worrying about its existence.

In April 1793, the Committee of Public Safety was established, an emergency measure to focus the war effort and restore order on the home front. For all intents and purposes, the Committee of Public Safety became the executive branch of the government, which needed a new head now that the old one had been severed. Unfortunately, it would grow into something more tyrannical than Louis XVI had ever dreamed of being.

The political atmosphere was perilous, for those in the elite above all. Denying impatient politicians made little sense, so the savants buckled and handed over the provisional meter. At the same time, the Academy laid out the new measurement system. In structure, it was identical to Jefferson's plan—just as the Virginian had derived the bushel from the foot, the meter would be cubed to create a standard of capacity, which in turn would define a standard of weight, with decimal multiples and submultiples providing the other measures. The big question was what all these measures would be called.

Many assumed that other than the meter itself, the old names of the measures—*palme*, *doigt*, *livre*, *pinte*—would simply be applied to the new system, again à la Jefferson. But certain savants wanted original names, fearing not only confusion between old and new measures but believing that novel and precise words could banish muddled, nonrational thinking. Such theories had led to a long-standing vogue for inventing "philosophical" languages in which each word had one precise meaning, and they inspired Lavoisier and his fellow chemists to abandon words tinged with alchemical overtones for such neologisms as *oxygen*, *hydrogen*, *carbon*, and *nitrate*. If people accepted the new language, they accepted their new theories. Outside France, the practice was often abhorred; Jefferson called Lavoisier's nomenclature "premature, insufficient, and false," while others detested the terms of thinly Latinized French.

Split between choices, the Academy presented two naming schemes to the government, one with the old names and another with all-new terms like *cadil* (the future liter), *grave* (kilogram), and *franc*, plus an utterly novel twist.

For subdividing the standards, the Academy proposed using the Latin terms for tenths, hundredths, and thousandths—*decima, centesima,* and *millesima.* These had been the inspiration for the dimes, cents, and mills that were used to divide the dollar, but could be applied to any measure; Jefferson would later use them to split the mile. The novelty of the Academy proposal was that it would employ the Latin roots as prefixes, attaching *deci-, centi-,* and *milli-* to the new names of the standards. The brilliance of this was in eliminating the kind of confusion where sixteen ounces can mean both a pound and a pint.

This radical scheme had its drawbacks, however. Compound names like *millimeter* and *decicadil* not only sounded too foreign but ran four syllables long. Why cast aside strong, monosyllabic French words in favor of lengthy, tongue-tripping pseudoclassical constructions? The decimal math was hard enough, and at least the old names were understood when it came to which were bigger and which were smaller.

If the members of the Academy were divided over which system of names to favor, it was quite clear which one the Jacobins preferred. Far from being a positive, the old names being easier to pronounce and more familiar was considered a detriment. Language, if left alone, kept alive ancien régime habits and feudal thinking, whereas new terms could be used as a tool to smash old ways and beliefs, and provide an education— or reeducation—of the citizenry.

That corrected language could lead to corrected thinking was an idea Jacobin politicians had taken to with a vengeance. As with the guillotine, language could level the differences between the old estates. Peasants were no longer required to address *monsieurs* with the deferential *vous;* now all were *citoyens* and no one too good for the familiar *tu.* The old provinces had given way to rational divisions with natural names shorn of history, and the same thing needed to happen with measures—their newly logical shapes had to be matched by names that were no less revolutionary.

On August 1, 1793, the decimal metric system with radical nomenclature and prefixes was officially passed into law, and the country was given eleventh months to begin its obligatory use. But no matter how much they liked the new measurement system, the Jacobins didn't care for the institution in which it had been incubated.

A day of reckoning was at hand for the Academy. It followed the death of its enemy Marat, who on July 13 was killed in his bathtub by the beautiful twenty-four-year-old assassin Charlotte Corday. His death gave no relief to his foes, as Marat became the martyr of the radical wing of the Jacobins, whose leader, Robespierre—the Incorruptible One—was shortly elected to the organ that would power his coming dictatorship, the Committee of Public Safety.

Any organization deemed "unrepublican" was in peril. The man doing the most to save the Academy of Sciences was also the one who had been pushing the meter project the hardest, Lavoisier. The chemist had become treasurer of the Academy two years earlier, and with its funding cut off he was paying its salaries out of his own pocket. With Condorcet in hiding—a warrant for his arrest had been issued shortly before Marat's murder—Lavoisier was effectively running the Academy.

On August 8, a bill that proposed the closing of all academies was presented to the Convention, and the savants were down to their last hope, the Abbé Henri Grégoire. Grégoire had impeccable credentials as a champion of the people, having been the first priest to cross to the Third Estate during the Estates-General. His main cause was the elimination of the divide between privileged Paris and the rest of France, which he believed could be achieved through universalism and uniformity. To that end, Grégoire would spearhead the movement for eliminating dialects in favor of one national language and also become a strong supporter of the meter project. Citing the metric system specifically, Grégoire argued that the Academy of Sciences was a positive force for the people and alone among all the academies should be spared.

Rising to speak against Grégoire was Jacques-Louis David, who had gone from being a highly paid portraitist of the wealthy to propagandist of the Revolution. At the time, David was at work on another of his masterpieces, *The Death of Marat*, and the painter channeled his deceased subject when he thundered, "Let us destroy—let us annihilate—these deadly academies, which can no longer survive under a free regime. Academician though I am, I have done my duty."

David's self-immolation had its desired effect, and his contention that all academies should be closed won the day, making the passing of the metric system into law one week earlier the swan song of the Academy of Sciences.

Revolutionary Times

The new system of weights and measures may not have been what the people of France had asked for, but it did address concerns they had frequently brought up. The same could not be said, however, for another area of measure coming in for similar treatment.

The decimalization of time had been in the offing from the start of the Academy's discussions. There was certainly as much reason to fix the clock as there was to fix the foot and pound, considering the awkward division of the day into halves, twelfths, sixtieths, and more sixtieths. There were also the fields of geometry, geography, and geodesy, which—though of no consequence to the common folk—mattered greatly to savants, some of whom had taken it upon themselves to begin using a 400° circle divided into hundredths.

In the spring of 1793, a division of the day into ten decimal hours of 100 decimal minutes had been proposed along with the rest of the metric system. The new *heure decimale* was nearly two and a half times longer than the old hour, noon occurred at 5:00, four in the afternoon at 6:67, and the stroke of midnight moved the clock not from 11:59 P.M. to 12:00 A.M. but 9:99 to 0:00. The main benefit of the scheme was knowing what percentage of the day had passed—50 percent, 66.7 percent, or 99.9 percent. Although gung-ho savants such as Lavoisier got decimal timepieces and pro-Jacobin towns kept ten-hour municipal clocks, the idea didn't catch fire.

The metric clock, however, was a sideshow compared to the reorganization of a far more convoluted apparatus, the calendar. At least every day was made up of the same number of hours and minutes; months had between 28 and 31 days and possessed no fixed relationship to the weeks at all. What could be less uniform and more irrational than that? But the reform of the months and weeks was less scientific than political, as republicans wanted a new calendar to mark the new era in the history of mankind they had brought about.

Five weeks after the suppression of the academies, a commission to overhaul timekeeping began to deliberate. A few days later, on September 17, the Law of Suspects was passed, an event sometimes marked as the official start of the Reign of Terror.

Despite its name, the Terror was not a time of random violence—that had been happening for years—but rather of repressive order handed down from the top, which increasingly meant Robespierre. The Law of Suspects would be the main legal instrument of the Terror, allowing mass roundups of entire categories of people to be arrested, with the most popular sentence handed down always seeming to be the same—death.

Imagine if the radical wing of one of today's political parties were given a working guillotine to selectively use against the other, and you might have an idea of how things went down. Opposing politicians, priests who had actively upheld their vows, and Marie Antoinette all joined what became a parade to the National Razor.

Against the backdrop of the Jacobin consolidation, a Temporary Commission of Weights and Measures was appointed. Mixed in with bureaucrats were the savants, demonstrating that the meter project had indeed provided a lifeline to those who might otherwise have been expendable. Lavoisier, Laplace, and other former members of the Academy were able to continue their work, but not the man who had for so long been its secretary. Already in hiding, Condorcet was condemned to death in absentia.

Though he was seemingly well positioned in post-Lafayette France, the days had turned too radical for even the prophet of progress. Not supporting the execution of the king and the company he had kept—his old monarchist friends who had fled or been killed—were both marks against Condorcet, but what doomed the former marquis was his blistering criticism of the new Jacobin constitution.

As the Terror began to ramp up, the new Revolutionary calendar came into form. It would keep the year divided into twelve months (the only twelve that escaped decimation), but little else would stay the same, starting with the fact that each month would consist of three *decades* of ten days each.

The replacement of the seven-day week with the ten-day *decade* would be as unmooring as any metric change, which was exactly the point. The motives behind the overhaul of the calendar had only superficially to do with decimals and everything to do with priests getting guillotined and a new deist religion Robespierre was concocting. In early November, Notre Dame was reconsecrated as the Temple of Reason, one more step in replacing the old and corrupt with the new and ideal.

The new calendar was meant to be another tool in the reeducation of the people; in this, its power to influence was unrivaled, as nothing controlled the daily lives of the people more. The calendar had traditionally been the domain of the Catholic Church, who used it to tell people when to work, and more important, when not to. Every day of the year celebrated a saint (if not two or three), with tales of their holy deeds dictating lessons that helped form a controlling narrative of people's lives.

The key to the revolutionary calendar was its eradication of Sunday. Whereas the Lord rested on the seventh day, the citizen would skip work on the tenth, called *décadi*. This made it impossible to observe the Sabbath more than one time out of ten times without running afoul of the law. The truest hope of the radicals was that once the new rhythm took over, people would forget when Sunday even was.

The names of the months could not be allowed to stand, either, fashioned as they were after gods (like Mars), tyrants (the Caesars, Julius and Augustus) and misplaced numbers (the ninth through twelfth months having been named after the Latin words for seven through ten—*septem, octo, novem, decem*). Here the Jacobins decided to get poetic, and they turned to a playwright, actor, and poet named Fabré d'Eglantine, who also happened to be a radical political figure and Danton's right-hand man.

The names d'Eglantine coined tied the months to nature and sounded distinctly French. Whereas the rest of the metric system used prefixes to order sizes, suffixes classified the months by season, with -*aire* indicating fall, -*ôse* winter, -*al* spring, and -*dor* summer. The autumn months were Vendémiaire (after the word for "vintage"), Brumaire (fog), and Frimaire (frost); the winter ones Nivôse (snow), Pluviôse (rain), and Ventôse (wind); spring Germinal (sprout), Floréal (flower), and Prairial (field); and summer Messidor (harvest), Thermidor (heat), and Fructidor (fruit).

Christian holidays were supplanted by Republican ones, with a long festival to end the year that used up the five or six extra days orphaned by the twelve 30-day months. These holidays were to be called the *Sanculottides*, in honor of the urban workers on whose backs Danton and Robespierre had ridden to power. Instead of every day commemorating the life of a saint, each was assigned a vegetable, animal, tool, or some other element of nature or agriculture to honor. So 19 Nivôse celebrated the radish, 15 Fructidor the trout, and 20 Floréal the hoe.

The calendar also needed a new start date, both in historical terms— *Anno Domini*, in the year of the lord, was unacceptable—and time of year. By a stroke of luck so remarkable it seemed fated, the republic had been declared on the autumnal equinox, in which day and night are equal. What could be more ideal for marking the start of the era in which all men had become equal?*

Along with the rest of the new calendar, the new start point of time— September 22, 1792—was declared on November 24, 1793, a date which instantly became 4 Frimaire Year II. That same day, the revolutionary government decreed the arrest of all former shareholders of the General Farm. The bloodsuckers were an easy target, seeing as their riches had come from collecting taxes and their still-fat bank accounts could be confiscated to fund the Revolutionary cause. That Lavoisier was arguably the most indispensable member of the measurement commission didn't save him, although he made the case that it should.

The worst had only just begun.

The Final Decimation

In the Anglo-Saxon world, the face of the Revolution would long remain the Terror, which never lost its ability to fascinate, repulse, and amuse. And nothing amused or proved quite so ready for ridicule as the new calendar, the months of which one British wit rendered as Wheezy, Sneezy, Freezy; Slippy, Drippy, Nippy; Showery, Flowery, Bowery; and Hoppy, Croppy, Poppy.

The bloody rule of the Jacobins made prophets of those Americans who from the start had been predicting the Revolution would turn out

---

* Starting the year on a solstice or equinox had precedent in many cultures, including the Catholic one. The conception and birth of Christ had been tied in ancient times to the spring equinox and winter solstice—which is why the Annunciation is March 25 and Christmas December 25—and also the reason that the Christian calendar had customarily started in one or the other period. In America, a January 1 New Year's Day was only forty years old, and some European countries had yet to adopt it.

badly. "I know not what to make of a republic of thirty million atheists," John Adams had written in 1790, and he stated that tyranny by the majority could be as bad as that by any king.

Events had also changed the Franco-American political relationship. The United States was a formal ally of France, going back to its own Revolution and the aid it had received. Now that France was at war, Jefferson and others wanted the United States to return the favor. Opposition to this was led by Hamilton, who was no fan of the French Revolution and thought any involvement in the affair would be ruinous.

The dispute was escalating as Washington began his second term, which he hadn't wanted but felt compelled to take lest his own country fall into factionalism. The month after his second inauguration in 1793, Washington declared American neutrality in the growing European conflict. Jefferson believed the president's issuing of a proclamation was going too far, and he saw the cause of France further dented in his home country by the machinations of Edmond-Charles Genêt, the French minister to the United States, who was commissioning American privateers to attack British ships. The American government angrily demanded the recall of "Citizen Genêt," but with the Jacobin rise to power, that started to look like a death sentence, so Washington granted Genêt asylum in the United States, where he would stay for the rest of his life.

The brightest star of the former Academy of Sciences had no such options. In what was in other places the final month of 1793, the savants on the temporary weights and measures commission tried to rescue Lavoisier, seconding his claim that his nearly finished work on establishing the weight of water was indispensable to the meter project. Not only was the request denied, but Robespierre's Committee of Public Safety purged most of the remaining ex-Academicians from the measurement commission, which itself was dissolved come springtime, the decision having been made that the provisional meter was good enough as it was. There was no more need of surveys, or of savants.

While still in hiding, Condorcet wrote his signature work, *Sketch of an Historical Picture of the Progress of the Human Mind*, which was—rather implausibly, given his situation—a tract of utter optimism. Traveling under an assumed name, Condorcet was caught in late March 1794 and placed in the prison cell where he died, probably of poison, likely his own. He saved the guillotine some work, but plenty of others didn't. Danton soon met the National Razor alongside his ally d'Eglantine, whose fall from namer of the months to enemy of the state was remarkably swift. On May 8 came the turn of Lavoisier. The greatest scientist of the age and the father of chemistry, Lavoisier could not escape that he was also a tax farmer. No shareholder of the General Farm who remained in France survived the Terror, but Lavoisier was not just anyone.

Amid this orgy of death came the Festival of the Supreme Being, a further step in the created religion of Robespierre. It was the apex of the Terror, and against this backdrop of execution and ersatz religion, the first truly superfluous and nonsensical change of measurement standards was made. On 12 Germinal—April Fool's Day, by Gregorian reckoning—the Centigrade scale was adopted as the official measure of French thermometers.

Numbers had first been put to temperature in the early 1600s by the sublimely named Santorio Santorio*; prior to that, weather could only be described in adjectives. By the end of the seventeenth century, there seemed to be as many scales as instrument makers, with a bewildering

---

* If anyone in the history of measurement deserves a footnote, it is Santorio Santorio, a pioneer in making the Age of Science also the age of measurement. Santorio was the first to come up with tools to measure the human pulse, humidity, waves, and more. In his best-known experiment, he recorded the weight of everything he ate, drank, and excreted—as well as himself at regular intervals. The numbers didn't square, and the missing weight he attributed to "insensible perspiration." For his work in this field he has been called the father of metabolism.

variety of degrees. Thermometers began to come with cards marked in more than a dozen different scales, but eventually the Anglo-Saxon world coalesced around the Fahrenheit standard and continental Europe the Réaumur scale, named after the member of the French Academy who devised it in the 1730s.

Alone among metric changes, the move to Centigrade did not in any way assist math or calculation, nor did it erase some bad habit of the ancien régime. The adoption of the Centigrade scale had nothing to do with logic, and everything to do with aesthetic purity. Centigrade gave the appearance of being more decimal than the Réaumur scale it replaced by setting the freezing and boiling points of water to 0° and 100°—but only the appearance. The inch was duodecimal because twelve of them made a foot and the centimeter decimal because one hundred of them made a meter, but 100° makes 100° and nothing else, whether it be in degrees Centigrade, Fahrenheit, or Reaumur. Thermometers did have a great failing in logic, which was that their scales contained negative numbers, a mistake Centigrade only reinforced.

Having changed distance, weight, capacity, coinage, the clock, the calendar, the circle, the earth, and now the thermometer, the French Republic had finally run out of things to decimate. The people, meanwhile, had run out of patience with change itself. At the beginning of the Revolution, there had been little desire to get rid of the king, only to fix the finances of the state; few had wanted to annihilate the Church, only to curb its abuses; and people hadn't asked to be rid of pounds and feet, only that existing measures be fair.

In the middle of the summer of Year II, the man responsible for so much excess met his fate. During the night of 9 Thermidor, Robespierre and his cohorts on the Committee of Public Safety were declared outlaws and arrested, though not before the Incorruptible One blew apart his jaw in a botched suicide attempt. He was dispatched to the guillotine without trial the next day, the victim of the same brand of justice he had deployed so frequently.

In the days that followed, associates of Robespierre and his circle were rounded up and detained. Placed briefly under arrest and incarcerated was a young Corsican army officer in whose future hands the fate of the metric system would lie, as would so much else.

6/16

or Three Eighths

NAPOLEONIC MEASURES

T HE ADDRESS of General Washington to the People of America
on His Declining the Presidency of the United States appeared
in newspapers and pamphlets in the fall of 1796. In it, the soon-to-be-
former president advised his fellow countrymen to avoid entanglements
abroad and factionalism at home. These warnings have been employed
for all manner of arguments since, but at the time Washington was
targeting the political infighting that continued to rage over the situation in Europe.

The most remarkable thing about what would come to be called
Washington's Farewell Address was its very existence. Beyond the
democratic spectacle of a head of state speaking directly to the people
was the fact that the leader had been able to make it through eight
years. Most unusual of all was that he was turning down the opportunity to perpetuate himself in power.

The health of the American revolutionary experiment stood in stark
contrast with the one in France, where a third constitution in four years
had just been adopted. During Washington's two terms, France had been
ruled by the king, National Assembly, Constituent Assembly, Legislative
Assembly, National Convention, Committee of Public Safety, and
Directory, which doesn't even begin to tell the tale of the individuals who
actually wielded the greatest political power at any given time.

The French situation stabilized somewhat with the Thermidorean reaction that followed the fall of Robespierre, a period that saw the first veering back to the right after so many years of leftward lurch. In putting together the 1795 Constitution, those in charge attempted to undo the dangerous excesses of the Jacobins, such as giving everyone the vote. *Vous* and *monsieur* made their comeback as polite forms of address, as did wearing fine expensive clothes out on the boulevard. The Cult of the Supreme Being was dumped, and many imagined that the strange and incomprehensible new measurement system would follow it, particularly after the meter's pitiful introduction in the last summer *decades* of Robespierre's tyranny, when—despite becoming obligatory—it remained invisible, with less than one thousand meter sticks having been produced.

They would be disappointed. Far from being abandoned, the metric system was reinforced, a sign that France's new masters had no desire to undo the Revolution. Certainly, they wanted nothing to do with kings, having suppressed royalist riots in 1795, most famously with the "whiff of grapeshot" fired into a crowd by order of a young General Napoleon Bonaparte. The only part of the metric system that followed Robespierre's pseudoreligion onto the scrap heap was the *heure decimale*, though it would linger a few years in Jacobin strongholds and on some public clocks. The official explanation was that decimal time was of no use to ordinary citizens and "only benefits scientists." The calendar survived, with one major complaint against it—the lack of days off—rectified by adding a midweek break. Citizens who resented getting only every tenth day off now found themselves with more days off than ever. Improvements were also in store for weights and measures.

The metric system as we know it came into being with the law passed 18 Germinal Year III (April 7, 1795). New to this version was the introduction of a second line of prefixes and the declaration of the liter and kilogram as the principal standards derived from the meter.

The original prefixes had handled division and come from Latin, while the new ones were for multiples and came from Greek. There was the *kilo-* (x 1000), the less familiar *hecto-* (x 100) and *deca-* (x 10), and the now defunct *myria-* (x 10,000). The new prefixes allowed there to be a single root name for each category of measure, whereas previously there had been multiple ones, for example with weight, where the *bar, grave,* and *gravet* or *gram* were rough equivalents to the *ton, pound,* and *grain.* The *grave* was the standard, but its name was now dismissed for the bizarre reason that it was a homonym in German of *Graf,* a title meaning "Count," and so deemed inconsistent with republican virtues. Instead, *gram* became the root name, which in turn made *kilogram* the name of the standard, albeit awkwardly.

Also introduced definitively was the *franc.* Originally, Academy savants had planned to base the new currency on the metric weight of precious metal, which suggested a ten-gram coin of silver as a standard. Political expediency won out over decimal purity, however, and the new measure was effectively the old *livre* in disguise, rounded to its nearest metric equivalent, five grams of silver. For similar reasons, it was not given a new and unique name, but called the *franc,* which was already popular slang for *livre.*

The weight of the franc, however, remained purely theoretical. The first francs contained no precious metal at all, but were made of paper. These bank notes were called *assignats,* and were similar to what American *continentals* had been. Under the Terror, a citizen didn't accept *assignats* at the risk of his head, an effective anti-inflationary policy, but the lack of such a threat after Thermidor led to catastrophe. By the summer of 1795, the *assignat* was worth less than a centime on the franc, a trend that would soon reach dizzying proportions and give the world its first case of hyperinflation.

An all-around brighter situation was to be found on the battlefield. War, it turned out, was one thing that a revolutionary republic was very, very good at.

THE METRIC SYSTEM OF 1795

1793/94 unit	1795 unit	Value
Length		
quadrant (of meridian)	—	10,000,000m (10,000km)
grade	—	100,000m (100km)
—	myriameter	10,000m
milliare	kilometer	1000m
—	hectometer	100m
—	decameter	10m
meter	meter	1m
decimeter	decimeter	.1m
centimeter	centimeter	.01m
millimeter	millimeter	.001m
Area		
are	hectare	10,000 sq m
deciare	—	1000 sq m
centiare	are	100 sq m
—	centiare	1 sq m
Volume		
—	decastere	10,000 cu dm
cade	kiloliter/stere	1000 cu dm (1 cu m)
decicade	hectoliter/decistere	100 cu dm
centicade	decaliter	10 cu dm
cadil/pinte	liter	1 cu dm
—	deciliter	.1 cu dm
—	centiliter	.01 cu dm
—	milliliter	.001 cu dm
Weight		
bar	—	1,000,000g (1000kg)
decibar	—	100,000g
centibar	myriagram	10,000g
grave	kilogram	1000g
decigrave	hectogram	100g
centigrave	decagram	10g
gravet/gram	gram	1g
decigravet	decigram	.1g
centigravet	centigram	.01g
milligravet	milligram	.001g
Coinage		
	franc	1 franc (5g silver)
	decime	.1 franc
	centime	.01 franc

The law of 18 Germinal Year III (April 7, 1795) established the near-definitive metric system. The move to having a single base name for each series caused much rejiggering, most awkwardly with weight, where the base unit (gram) is different from the standard (kilogram). The base name of area (are) was demoted, making the hectare the nearest equivalent to the acre, while in capacity measures the cadil of 1 cubic decimeter was renamed the liter and the cade of 1 cubic meter became the stere. The stere, however, was meant for dry substances, particularly firewood.

Long-term success began in the final months of the Terror, when French troops expelled Austria from its portion of the Netherlands (present-day Belgium) and drove them as far as the Rhine. The following year saw even greater French success in the Dutch Netherlands, which had recently come under the heel of Prussian troops crushing the Patriot Revolt, a movement inspired by the American Revolution.

These foreign troops were no match for the secret weapon of the French Republic, the citizen-soldier. France had a massive population and was the first country to mobilize a national army via something like a draft. The occupying armed forces of alien despots and mercenary Hessians were no match for the many peasant sons of liberty.

In no part of the Low Countries were the French entirely unwelcome, either. The Netherlands had a thriving liberal culture, with the *Leiden Gazette* having long been the European newspaper of record and the "only one worth reading," according to Jefferson. It had drummed up European support for the American Revolution and was the source many Frenchmen turned to before 1789 for news suppressed in their own country. Many Dutch felt they had played a part in making the French Republic and wanted something like it for themselves. A Batavian Republic was established in advance of oncoming French troops in 1795, which the *Leiden Gazette* proclaimed year one of Batavian Liberty on a masthead that now held the motto *Liberté, égalité, fraternité.*

Similar governments cropped up nearly everywhere French armies marched over the next few years, with those who formed the republics believing themselves to be patriotic democrats, and those resisting the French considering them traitorous collaborators. Seeking to export the principles of the Revolution, the French cultivated these sister states, with Napoleon Bonaparte going on a veritable binge of republic making after spectacularly driving the Austrians out of northern Italy.

Like the Netherlands, Italy contained both a native population sick of foreign domination and a liberal advance guard. This was especially true in Milan, which Napoleon entered triumphantly in the spring of 1796. Milan was the home city of Cesare Beccaria, who had died a year and half earlier but remained one of the most influential thinkers of the age, particularly for the liberal penal code he had devised and his opposition to the death penalty.

When it came to measurement, Beccaria was also in the vanguard, believing in the decimalization of weights, measures, and coinage and looking to a portion of a meridian as their basis. In the 1770s, he was asked by the Austrian rulers of Milan to reform weights and measures in their Italian possessions, and so at the same time his friend Condorcet was trying and failing to give France a uniform, scientifically based system of measurement, Beccaria was succeeding in doing just that for Lombardy. His reforms became the immediate model for the metric system, which in turn would be forced on Italy and threaten what Beccaria had created.

Condorcet had claimed the meter to be *for all time, for all peoples*, and the push to make that happen was beginning. To this end, French savants were finally ready to cast the provisional meter aside and proclaim the official one. They would do so at the first international scientific conference ever held.

International Savants

Savants were just one more segment of the elite who had made a comeback in post-Jacobin France. The same official who had been responsible for their purging in 1793 praised the *philosophes* as he invited them to take back charge of the meter project and revived meridian survey, while the academies were reconstituted as the Institut de France.

No savant had come through the Terror quite so well as Pierre-Simon Laplace. With Lavoisier gone, Laplace became the greatest name in French science, if not its most beloved person.

When it came to the metric system, Laplace was a true believer. A member of the original two metric committees, he had also been a source of controversy. Some had blamed him for the boondoggle of the meridian survey, saying the physicist needed its data to finish the work that would become *Celestial Mechanics*, the first draft of which he finished in 1798 and would cement Laplace's reputation as France's Newton.

Laplace was particularly well positioned in the new France, having a link to the man of the moment. While a sixteen-year-old student at the École Militaire, Napoleon had Laplace as his examiner in math, which is something like taking an oral report with Shakespeare or having your science project graded by Einstein. Napoleon was training in the artillery, which involved the study of mathematics and was at least part of the reason the Corsican fancied himself a man of science. His election to the Institute in late 1797 pleased him greatly, however much it was a gesture of flattery.

By that time, the Treaty of Campo Formio had been signed, cementing Napoleon's advances in Italy and burying the Austrian-led First Coalition. France's only remaining opponent was Britain, which would attempt to thwart Napoleon's invasion of Egypt. As part of the Mediterranean expedition that departed in spring 1798, Napoleon included more than 160 savants, whose research would help lead to a mania for all things Egyptian that lasted decades. Discoveries such as the Rosetta stone would be fundamental in the field of linguistics, while archaeology benefited from a new survey of the Great Pyramid, the results of which would show (some savants believed) its dimensions to be based on a portion of the earth's circumference, just like the meter.

Though otherwise trying to ingratiate himself with his former student, Laplace declined Napoleon's invitation to Egypt, having a different plan in mind. Rather than send the savants of France abroad, he wanted to bring the savants of the world to Paris. The ostensible reason was to get their aid in checking over the results of the meridian survey. Once that was done, the circumference of the earth could be calculated, the result divided by forty million, and the official meter declared.

The French had long hoped that America would be the first country to voluntarily adopt the metric system. At the suggestion of the Abbé Grégoire, a metric emissary had been sent to the United States in early 1794 armed with copper replicas of the meter and grave, but he died en route. A year later, President Washington asked Congress to look into the metric system, but nothing more came of it than his other exhortations or of Jefferson's proposed reforms.

Since then, the U.S. relationship with France had become only more troubled, in part due to the first contested presidential election in American history, which saw the Anglophile John Adams nudge out the Francophile Thomas Jefferson by three votes in the electoral college.

In his first months in office, Adams sent a diplomatic delegation to Paris to try to get France to stop seizing and looting American ships, which the French were doing on the grounds that American trade helped the British cause. The arriving American envoys, however, were seen as an economic opportunity by one French official, the venal Talleyrand. Talleyrand had waited out the guillotine years in exile in America and England, and upon returning home he was installed by the Directory to the powerful position of Foreign Minister. Through intermediaries—code-named X, Y, and Z—Talleyrand demanded of the Americans both a ten-million-dollar-plus loan to the French state and a $250,000 "sweetener" to himself, followed up by the threat of war when his terms were not met.

The XYZ Affair outraged Americans, unused to the bribery diplomacy practiced on the continent. "Millions for defense, but not one cent for tribute!" was the line made famous for generations. The row initially provided a huge boost to Adams and led to what became known as the Quasi-War. But most of all, the brazenness of Talleyrand demonstrated the changed position and mindset of France, both of which were sending shivers down the spines of its sister republics.

Even as more republics sprang up—the Helvetic (Swiss) and Roman ones in 1798, to mention just two—it was clear that these were not so much sisters as satellites, and their governments junior partners in their own affairs, with constitutions full of ideals but subservient to French interests. They were still allegedly homegrown, but nearly anyone could declare a republic and expect French support to make it stick. The Helvetic Republic was born with less-than-broad-based Swiss support, while the Roman one led to the pope's being taken as a prisoner to France, where he died in 1799. Not only were reactionaries horrified, but so, too, increasingly were local Jacobins.

Particularly galling were the massive indemnities the French were charging states for their "liberation," which helped to supply its coffers with the precious metal needed to make the franc more than a paper reality. While banks were being emptied in Amsterdam and Zurich, art treasures were getting sucked out of Italy in order to restore France to what Napoleon and others believed was her naturally supreme place in the arts, a position that had been lost with the looting and vandalism of the Revolution.

Not that Napoleon was shy about any of it. "I will prove another Attila to Venice!" he declared, and he made good on his promise, invading the world's oldest would-be republic and handing it over to the Austrian emperor after having stripped it of what treasures he liked, an act of thuggery that appalled even Talleyrand, which is saying something.

It was Talleyrand who sent out the invites to Laplace's international conference of savants, a gathering that looks rather less international when one realizes that its participants all came from either French client states or neutral countries cowed by its aggression. Not a single British, Prussian, Russian, Swedish, or Austrian savant came to Paris.

For Laplace, it hardly mattered. His desire for the conference hardly grew out of a sense of true collaboration, as he would publicly have it. "You realize that all this is only a formality," Laplace wrote to a member of the survey team who feared foreign scrutiny, "to enable them to consider the system on their own, to do away with national envy, and to make them adopt the measures."

The only savant from a neutral state to attend the conference, the Danish astronomer Thomas Bugge, felt he was being used by his hosts and left. Bugge also believed there to be fatal flaws in the meter project. No matter how precise the survey of the meridian had been, he pointed out, the idea that it could be definitive was foolhardy, as was the notion that the meridian-derived meter could be said to be taken from nature.

The members of the commission who stuck around, however, went about their task enthusiastically. The scientific headline from the conference was that the earth was more squashed than previously realized, and—even more important—wasn't squashed evenly. It was bumpy. This meant that a measure of the French meridian was a measure of *only* that meridian, and the meter taken from it a portion of France and nowhere else.

As for that meridian measure, it resulted in a meter that was only a hundredth of an inch different from the provisional one based on the meridian survey of 1740. What's more, it changed the meter in the wrong direction. The author of the definitive book on the meridian survey, Ken Alder, writes, "Seven years of labor had only succeeded in making the meter less accurate." But even at the time the minuteness of the difference led to scorn. "It should not have been necessary to go so far to find that which lay so near." Critics of the project had been right from the start; ultimately, the survey had sprung from the ulterior motives of savants and not been at all necessary to developing a system of measurement.

Despite the difference of little more than a hundredth of an inch, the provisional meter was replaced, changing the measures of France again. A physical meter was forged out of platinum to be the official standard, and deposited in the Archives on 4 Messidor Year VII (June 22, 1799). In his speech at the presentation ceremony, Laplace said that with the meter the poorest farmer could now say, "The field that nourishes my children is a known portion of the globe. In that proportion I am a co-owner of the world."

Once the show was over, the foreign savants left with iron copies of the meter in their gift bags and the express wish of their hosts that they put them to use back home. They had little choice.

Napoleonic Fixing

If the metric system was adopted abroad, it would be the first place it was adopted.

In France, steps had been taken to try to educate the people in the new system, with publications distributed that ranged from conversion charts to pamphlets to manuals. Playing cards, already purged of royalty—the jack, queen, king having been deposed by figures depicting *liberté, égalité, fraternité*—were used ace through ten to teach the decimal system. Physical prototypes of standards were bolted to prominent walls for public display, the same as the old standards had been, and graphic posters endeavored to visually explain the new measures.

None of it mattered. It's one thing to use wordless pictures to instruct the masses on how to put together an IKEA bookshelf, a whole other to teach a new way of measuring to an illiterate, innumerate populace. Public opinion remained hard against the new math and long foreign-sounding names with the bewildering prefixes. "Decimeter" and "decameter" sounded so similar, yet one was three and a half inches and the other thirty feet. Where people found the metric measures unavoidable, they simply converted them into the old measures they knew.

To savants, the fault lay not with the metric system but the people who refused to accept it. Any criticism—such as suggesting that the prefix system was too complicated for the average citizen—was met with harsh rebuke. "You cannot attack a part of the system without endangering the whole," said one savant at the new agency of weights and measures. Considering that the most educated minds in the nation had endeavored at such length to put together the metric system, its "goodness" should not be doubted nor its wisdom be attacked.

The government was also finding it difficult to stomach the opinions and behavior of their countrymen. Despite a new constitution aimed at curtailing the populist tendencies of the electorate, the people still weren't voting the way the ones in charge wanted them to, so the government kept tossing out election results, with the result that coups were becoming annual events.

Such a political situation was insupportable, and it was into this fray that Napoleon injected himself, returning to Paris after Nelson's annihilation of the French fleet in the Battle of the Nile. Backed by Talleyrand (who knew a winner when he saw one), Napoleon threw in with conspirators from within the Directory itself. Together they installed a three-headed dictatorship called the Consulate in a coup that took place on the night of 18 Brumaire (November 8, 1799). The next month, Napoleon betrayed his fellow plotters and installed himself as First Consul, effectively taking over the French government.

The elevation of Napoleon was good news for his old examiner Laplace, who presented him with a freshly published copy of *Celestial Mechanics*. The busiest man in the world joked that he would read it whenever he had a spare three months. Napoleon had Laplace and his wife to dinner, and the savant soon found himself appointed minister of the interior, one of the key positions in government and a perch from which he pushed the metric system.

It wouldn't last long. When Napoleon later said it was a mistake to involve mathematicians in administrative positions, he was surely thinking of Laplace, whom he considered a disaster in the job and replaced after three months. Napoleon, as it turned out, had little patience for savants and theorizers.

"We have finished the romance of the Revolution," Napoleon said shortly after his consolidation of power, "we must now begin its history, only seeking for what is real and practicable in the application of its principles, and not what is speculative and hypothetical. To follow any other course at the present day would be to philosophize and not to govern."

This approach was the secret to the phenomenally successful early years of Napoleon's reign, during which he drastically improved the efficiency of the French state. Within a year of the Brumaire coup, the metric system was again overhauled. Out went all the prefixes and nearly all the new names (the *meter* itself being the most prominent exception), and in came the old ones familiar to the people, who couldn't make out a kilogram from a hectogram but knew a *livre* was bigger than an *once*. Not that France went back to the old ways. The decimal system was retained, such that the *livre* was the kilogram in disguise, now more than twice its old weight, and the *once* was a tenth of a *livre*, not a sixteenth.

Pro-metric savants feared that dressing the metric measures in ancien régime names was the beginning of the end of their system, but Napoleon had simply undone the choice of the radical Jacobin days and returned the measures to the names most thought they should have had in the first place.

Napoleon had a more direct hand in creating the Code Civil, his proudest and most enduring achievement, which would eventually be given his name. The lack of a sound legal system had been one of the great failings of the Revolution, as idealists sought to re-create the law from scratch in the same way they had everything else. The Code Civil took from what was in place under the ancien régime—the Frankish-based *coutumes* of northern France and the Roman law dominating the south—and created a hybrid legal system shorn of feudal anachronism. It sought to make the law universal, clear, and easily contained between the covers of a simple-to-read manual. No more would there be a gnarled thicket of inconsistent customary practices to puzzle over. In seeking a rational, uniform system, the Code was law made metric.

The Code Civil would march alongside the meter, kilogram, calendar, and franc across Europe as part of a kit of rational administration that included other tools such as the department system and the Concordat, all wrapped up in a tricolor banner. Being spread no longer, however, were the ideas of republics, constitutions, and democracy, not even for show. And not even, much longer, for France.

The Spread of Empire

Even before Thomas Jefferson beat John Adams in the presidential rematch of 1800, he had worries about the new master of France. The Virginian would not have been sad to see some Brutus kill Napoleon and the other plotters of the Brumaire coup, but given that that didn't happen, Jefferson believed the Corsican might be the man best able to restore to France the "liberty, equality & representative government" he claimed to stand for.

Such delusions were soon abandoned. Jefferson would grow to detest Napoleon as "a cold-blooded, calculating, unprincipled Usurper" and one of the greatest monsters in history, as well as the man who more than any other destroyed the beautiful dream of republican France.

The nation's third president did, however, believe that Napoleon had done one great thing: sell France's vast colony of Louisiana to the United States. Particularly excellent was that he let it go at a fire sale price few in America could even believe—3¢ per acre. In one fell swoop, the size of America doubled.

The impetus for France's selling of Louisiana was its failure to squelch the Haitian Revolution. The lack of success overseas and in naval operations continued with the disastrous 1805 Battle of Trafalgar, where Admiral Nelson put an emphatic stamp on Britain's naval dominance. With Trafalgar, the oceans closed to Napoleon, and whereas the British were building their empire on the seas, Napoleon would build his across a continent. And an empire France had officially become.

That the grotesquely enormous 20' x 32' *The Coronation of Napoleon and Josephine* could have been painted by the arch-Jacobin Jacques-Louis David shows just how much the Revolution was having its *Four legs good, two legs better* moment. Or rather, that the Revolution had been extinguished.

As went France, so went the rest of Western Europe. States with ornately classicized names like the Batavian and Cisalpine Republics made way for the Kingdoms of Holland and Italy, with thrones occupied by members of Napoleon's family. But even that was rarely good enough for the emperor, who came to prefer outright annexation.

France had been nibbling at its neighbors since the first military victories of the Revolution, taking a piece of Spain here, a Swiss canton there, before graduating to larger bites like Belgium, the German lands west of the Rhine, and the sizable Italian state of Piedmont. These all were said to merely extend France to her "natural" borders. As emperor, however, Napoleon annexed the Kingdom of Holland in one fell swoop, meanwhile working his way down Italy in chunks, taking first Liguria, then Tuscany, and finally Rome.

As radical as anything Napoleon did to the map of Europe was his smashing of the nearly thousand-year-old Holy Roman Empire, a loose configuration of German states that included kingdoms, fairy-tale principalities, bishoprics, and free cities. There were well over three hundred *Kleinstaaten*, which doesn't even tell half the story, as virtually all of the "little states" had multiple noncontiguous parts, making the insanely jumbled map of *Mitteleuropa* the very picture of the irrationality of feudal ways.

In the end, only about forty German states survived. As many of them as could be corralled were placed into the Confederation of the Rhine, where they were kept under Napoleon's thumb and used as a breeding ground for the French army and a source of tax revenue.

All of the territories under imperial control were reorganized into departments and given what was now called the Napoleonic Code. The aim was to sweep away all the old customs-ridden anciens régimes and institute a uniform, rational administration across Europe. For whatever altruism was claimed and benefits did accrue, the goal was to create an empire that could be centrally controlled from Paris.

The metric measures were a part of this process. In the same way the Roman Empire had spread their twelves along with their laws, and Charlemagne his measures as a part of his *coutumes*, so Napoleon spread tens along with his code. Practical as ever, Napoleon allowed the metric measures to be translated into localized versions. So in Italy there was the *metro* instead of the meter, and the *lira* and *centesimo* for the franc and centime, just as the month of Ventôse (February–March) had become Ventoso to Italians, Windmonat to Rhinelanders, and Windmaand to the Dutch.

Imperial reforms were not equally successful everywhere. They went down best in places in the heartland of the empire, an area that radiated from the Rhine and included Milan, Zurich, Bavaria, Baden, and the Low Countries. This region saw an explosion in commerce from Napoleon's abolition of the Rhine's infamous tolls (collected in the Middle Ages by the original robber barons), and it was here that Napoleonic rule was most accepted.

The farther from the Rhine, the less easily reform went. Whereas feudalism had been gone from places like Amsterdam and Milan before French troops had ever marched in, other parts of Europe still had it in nearly pure forms. In general, Napoleon demanded that the legal code and metric system be adopted quickly, but even the emperor made allowances for places he deemed particularly backward.

Of the places looked down upon from Paris, high on the list was the south of France itself. The issue was particularly acute when it came to language. During the last month of the Terror in June 1794, the Abbé Grégoire had launched a national crusade to have the language of Paris supplant the *patois*, which dominated the south. Grégoire specifically linked this crusade to the metric one, wanting to make language similarly uniform and universal. All non-French languages and dialects came to be banned in government and in schools, while across the empire, French became the required tongue of administration, making it one more tool in the imperial kit.

The south of France was called the Midi, a word that, like "meridian," comes from the Latin *meridiem*, which means "midday" but also "south," because that is the direction the sun lies at noon in the Northern

Hemisphere. For the same reason, the Italian south was called the Mezzogiorno, a benighted area considered to begin in the Papal States. Not for nothing was the Catholic Church the enemy of Enlightenment, with the pope being literally against modern light, prohibiting gaslit street-lamps in his considerable temporal domain. He also forbade vaccinations, which the French administration introduced. Things were little better in Naples and the Kingdom of the Two Sicilies, which remained positively feudal, or heavily Catholic Spain, where the Inquisition still existed.

These Mediterranean lands would be among the places most resistant to French rule, and where meter sticks were all but unknown. Freedom fighters like Fra Diavolo (Brother Devil) helped impede the takeover of the Mezzogiorno for years. In Spain, resistance was even fiercer. Napoleon drove out the Bourbons and put on the throne his brother Joseph, who eliminated the Inquisition and divided Spain into departments, though they never went into effect. Constantly referred to as Napoleon's ulcer, Spain in this period gave the world the word *guerrilla* and Goya's nightmarish images of war.

Fighters in Spain warred against the "false philosophy" of the French, and "placing liberty above old customs." Far from being irrational, the customs were seen from within as the sole defense against the imperial monster. Uniformity might level the playing field, but it also smoothed the way for a universal dictatorship.

Across the empire, the franc was the greatest metric success, with mints from Rome to Amsterdam cranking out coins with Napoleon's profile on them. The calendar was a different story. In the early, idealistic days of the sister republics, it spread along with the tricolor and constitutions, but it was never much used by the people. It was employed mostly for administrative purposes, and it existed in places as far afield as Egypt and Louisiana (confounding genealogical researchers to this day). Napoleon himself scrupulously employed it while famously neglecting the rest of the metric system, demanding always to be spoken to in traditional units.

The calendar's death knell was its being irreconcilable by design with the Christian Sabbath. Legal worship of Sunday had returned with the Concordat that Napoleon signed with the pope, and shortly after, the Italian Republic dropped the Republican calendar. It had never been popular in France either, no matter how many extra days off, and Gregorian dating officially returned on New Year's Day 1806.

Laplace and other savants saved the decimal measures from following the ten-day week into oblivion. Over the next few years, the Empire redoubled efforts to make the metric system obligatory within its borders. So in areas that were German, Dutch, Italian, and Swiss, the metric system gained a foothold, although how much of one depended on the place. Strong local control in Geneva led to widespread metrication there, and the program also found success in the Belgian and Piedmontese departments. Northern Italy was a particular problem, on the other hand, partly because of the success of Beccaria's reforms, but it soon came under the imperial metric heel as well.

Metrication went only so far, however. Although used by prefects, savants, and customs inspectors, the metric system remained a flop at street level, a statement as true in Paris as anywhere else.

One's Place in the World

Since time immemorial the greatest issue of measures was that they were fair. "Ye shall have just balances," it says in the Old Testament. Standards were traditionally kept in temples, and severe penalties were meted out to anyone who kept false measures—those aimed at screwing the customer.

Although ancien régime measures had become illegal, merchants everywhere held on to their old scales. For good reason. What else could a stallkeeper in the market do when all his customers demanded to buy their goods by the old *livre* or *libbra* and not the decimal one? That the old measures had the same names as the decimal metric ones since Napoleon's 1800 overhaul made things even more confusing, not to mention a situation ripe for swindling.

Multiple measures created paranoia. Even when both sides in a transaction were honest, customers were convinced that prices had been rounded up, while sellers from greengrocers to prostitutes lamented they could no longer get the prices they used to, and so exchanges ended with one or both parties feeling they had gotten the short end of the stick.

Decimal reckoning had proved to be the obstacle that was impossible to overcome. Multiplication and division by doubling and halving were the most advanced math skills most of the population would ever have, and up till now, they had served them well.

As an example, take one of the most common purchases of the day, buying cloth. For hundreds of years, cloth had been sold across Europe by the *ell*, *aune*, or some other variation of the word for "elbow." If a customer needed a small amount of fabric—a half of a quarter ell, say—the storekeeper would measure out the ell to his body and fold it three times to mark an eighth. In Napoleonic Europe, that eighth of an ell was replaced by 0.125 meters. That the meter itself was a new standard was problem enough, but easy compared to understanding a number that represented a tenth plus two hundredths plus five thousandths. To someone for whom the formulation "half a quarter" was preferable to the overly abstract "eighth," this math was insurmountable. In practice such formulations were never attempted—the old ways ruled—but having such a chasm between official and common transactions put the average citizen at a disadvantage, the opposite of how decimals were supposed to make everyone equal.

Savants had weighed air, measured the distance to the moon, and calculated the speed of light, but none of that helped someone trying to buy salt. Laplace said that logarithms had given astronomers half their lives back, a statement that more than anything showed how much of their lives they spent making calculations. Decimals were chosen not only *by* savants but *for* savants; the metric system was constructed for the people only insofar as the savants had deemed it best for them.

Nothing better sums up the disconnect between savants and the common man than Laplace's speech at the meter ceremony, when he waxed poetic over the peasant who would get such immense satisfaction out of knowing what percentage of the planet his patch of land took up. Condescension aside—and Laplace talked down to everyone, his fellow savants most of all—what was supremely misguided in his statement was that the peasant had understood the size of his land *before* the meter came along, not after. Only a mathematical genius like Laplace could conceive of the size of the earth and fractions in the millionths; the farmer understood area in relation to his labor.

Similar to how distance measures had originated with the human body, land measures had come from what it took to work something. The German *morgen* represented the amount that a farmer with his oxen could plow in a morning, as did the English acre. In poor areas where beasts of burden were in short supply—rural France, for instance—bucket measures were the norm. A *pottle* in England, a *fanega* in Spain, and a *tunne* in Sweden were all measures of both capacity and area, connected by how much seed was needed to sow a given field. In vineyards, the number of man-days it took to harvest might be the measure, or in hayfields, mow-days. To the mind of one who worked the land, these were rational measures based on universal ideas, not the meter.

Decimals just weren't working. Savants lobbied again to save their system, with Laplace even appealing to the emperor's ego by suggesting that the metric system be renamed the Napoleonic measures, as the Code Civil had been. The emperor, however, could hardly want his name put on something so unpopular. Realizing what the savants did not, Napoleon pulled the plug on the decimal measures February 12, 1812.

Napoleon said the sole goal of his new law was to have a single system of weights and measures. Uniformity, and *basta*. This was a significant goal, especially considering the prerevolutionary feudal state of affairs in France and the fact that it was all the people had ever really wanted in the first place. The meter and kilogram would continue to underpin the system, but measures were rejiggered to be nearer their prerevolutionary (Parisian) values, and their old subdivisions brought back. So the pound was set to half a kilogram and the foot to one third of a meter, again broken into sixteen ounces and twelve inches respectively. These new-old measures were called the *système usuel*.

To savants, it was a travesty. It was their turn to wonder if it had been worth all the trouble, without the grand ideological goals, without the pure decimal math.

The *système usuel* measures were a part of Napoleon's efforts to put things in order on the home front before embarking on his next campaign of conquest, the disastrous invasion of Russia that led to his downfall. Finally gaining the upper hand after two decades, those countries who had borne the brunt of French occupation and war endeavored to turn back the clock. At restored courts across the Continent, men returned to wearing wigs and breeches while shedding the metric measures and Napoleonic francs for their old *punds* and *piastras*. Spain restored the Inquisition, and it was back to the Dark Ages in the re-created Papal States, where the pope dismantled artificial lighting and outlawed vaccinations. The situation was so bad that even those liberals such as Lafayette who had never supported Napoleon tried to bring him back. But Citizen Napoleon's return as the prodigal son of the Revolution was short-lived, as the Hundred Days term for his restoration indicates. All it did was harden the resolve of the victors at the Congress of Vienna to exterminate the revolutionary threat for good.

Talleyrand had been double-dealing with foreign powers for years—the reason for Napoleon's dressing down of him as a "shit in a silk stocking"—but it allowed him to now make a better deal for France than many thought the country deserved. The price was a restored Bourbon monarchy that would prove to be one of the most reactionary on the Continent. Gone was the tricolor, the Napoleonic Code morphed back into the Code Civil, and the new king outlawed public use of the metric measures.

Napoleon was exiled to the windswept Atlantic island of St. Helena, where he delivered his own postmortem on the metric system. Napoleon claimed to have always believed that the measures of Paris should simply have been applied to the entire country, as had been the common thinking in 1789. Weights and measures were an administrative question, he said, and involving mathematicians was an error that had doomed France to generations of unnecessary difficulty.

Savants like Laplace were the villains in Napoleon's estimation. It wasn't good enough for them to help forty million Frenchmen to better measures, they wanted to sign up "the whole universe," even though the pride of the British and Germans would never allow for it. Moreover, Napoleon said that the meter's enforcement against the will of the people was comparable to the lack of regard for the vanquished shown by some Greek or barbarian conqueror, and he expected that flaws in the meridian survey would soon be discovered. So what was the whole point in the end? "It is a tormenting of the people for mere trifles."

Others saw the matter differently, even in America. The man who had been the U.S. ambassador to Russia during Napoleon's invasion was positively besotted by the metric system, having written of it as the greatest invention since the printing press. By the time of Napoleon's harangues at St. Helena, he was America's secretary of state, and on his way to becoming the nation's next president.

7/16

LIGHTHOUSES OF THE SKY

O<small>N A</small> clear early morning in the heart of summer, with the just-risen Venus visible in the gathering light and the flame of the new Sandy Hook lighthouse still burning, the hills of Neversink came into view. It was the first time in eight years that John Quincy Adams had seen his native land.

Returning home gave Adams mixed feelings. He had set off for Europe with his father at age ten and turned fifty on the present voyage; in between, he had spent well over half his life abroad. As a boy, he twice joined his father on European missions, and he had served his country for the first time at age fourteen, as Francis Dana's translator at the court of the Russian czar. (Adams didn't get the job for a proficiency in Russian but in French, the language of European courts.)

Possessed of a thorny, thin-skinned personality, Adams was not a diplomat by disposition, but it was a life he had been born to. He served as U.S. ambassador to Prussia, the Netherlands, Russia, and finally Great Britain, an appointment he received following his brokering of the Treaty of Ghent.

Ghent concluded the War of 1812, America's peripheral slice of the Napoleonic wars. The treaty put the country not just at peace but at ease, as the United States had struggled with its place in the world since declaring independence, four decades earlier. The Era of Good Feelings commenced, a time of galloping prosperity and national unity that saw the collapse of the party system.

Adams's 1817 homecoming bore a remarkable symmetry to Thomas Jefferson's in 1789. Like Jefferson, Adams was leaving an ambassadorial post to become the secretary of state; also like the Virginian, he had been charged by Congress with preparing a report on weights and measures in response to a presidential request. This time the president was James Madison, a man Adams had befriended over chess matches in the Jefferson White House. Adams shared the same passion for measure as Madison and Jefferson, especially for measuring by pace. The younger Massachusite knew his own to be two feet six and 88/100 inches, which he used to record the length of his daily walks.

Adams's interest in measures had blossomed into an obsession while he was minister to Russia, a position that left him with lots of down-time. He studied Paucton's *Metrologie* as though it were the Bible, and he undertook his own investigations comparing Russian with English measures. This should have been a straightforward job, as Peter the Great had set the arshin (Russia's version of the cubit) to 28 English inches and also adopted the foot in the early eighteenth century, but Adams was unable ever to make a match. A national system of measurement did little good if a country did a shoddy job of maintaining and duplicating its standards.

The new impetus to America's putting its measurement house in order came in Madison's final presidential address to Congress. The architect of the Constitution, Madison urgently wanted Congress to use the power given it in Article I to set the weights and measures of the nation. Thirty years on, there was still not a single federal law pertaining to measurement, only regulations by individual states. Madison made note of Jefferson's report, saying that although it was now outdated, he still supported its central ideas, particularly the use of decimals.

Adams reached out to Jefferson immediately upon setting to work on his assignment. Jefferson and John Quincy's father were by then well into their storied late-in-life correspondence, which included the elder Adams writing in fond recall to his fellow ex-president of those European days shared with "our John," noting that the boy had seemed as much Jefferson's son as Adams's own.

In 1817, Jefferson was settled into his intellectually restless and financially strained retirement at Monticello. Born one of the wealthiest men in America, he was now up to his eyeballs in debt. He had just sold his books to the Library of Congress to replace those destroyed in the recent British sacking of Washington, but couldn't keep himself from buying more. In the entrance hall of Monticello, Jefferson's interest in measurement and gadgets was on display in the Great Clock, which Jefferson had designed to include the days of the week. Its weights were so long, however, that a hole had to be cut in the floor to accommodate their descent.

In opposition to John Quincy Adams's metric affection, the years had done nothing to lessen Jefferson's antipathy toward the system, which he maintained was the offspring of French nationalist ambition and not based on sound scientific principles besides. Though still favoring a pendulum solution, he had modified his views somewhat as regarded what was best for the United States. The decimal dollar, Jefferson believed, had succeeded because of the chaotic situation of currency in America. With weights and measures, there had already existed a system common to all the states, making new standards unnecessary. Not that he had given up on decimals. He favored dividing the foot, pound, and other measures the same way he had divided the dollar, and to that end he outfitted his carriage with an odometer that registered distances in miles, dimes, and cents, which he claimed were understood universally and perfectly among the people he happened upon while driving. It also chimed every tenth mile.

On the issue of whether it was best to stick to an old and rickety system of measurement or go the more radical decimal route, Jefferson recalled the ancient conundrum of lawgivers. "Shall we mould our citizens to the law," he wrote to Adams, "or the law to our citizens?"

Jefferson wished him well in his task, dearly hoping that the new secretary of state would succeed in rallying the nation to a better system of weights and measures.

The Adams Report

For however anxious the secretary of state was to begin his inquiry into weights and measures, other events stole his attention away from it, particularly the incursion into Spanish Florida by Andrew Jackson, the hero of the War of 1812. The action was an international incident that put the United States into conflict with the disintegrating Spanish empire and its champion, the Holy Alliance.

Composed of Prussia, Austria, and Russia, the Holy Alliance was as illiberal and reactionary a crew as you could get. Its mission was to oppose revolution and all it stood for, most especially the ditching of kings for democratic republics. This applied to rulers who lived an ocean apart from their subjects, and the dissolution of Spanish America into revolutionary nations drew their ire.

These republics had been established by Spanish American leaders such as Simón Bolívar during the Napoleonic takeover of Spain, when the Bourbon king had been booted off the throne and royal administration in the colonies broke down. With the restoration of Ferdinand VII, himself an archreactionary, the attention of the Spanish crown had turned to recovering control of their overseas possessions.

While some in the Monroe administration were horrified by Jackson's incendiary action, Adams opportunistically used it to pry an enormous swath of territory away from Spain, which was willing to deal away what had never been a crucial part of its empire in order to focus on more valuable possessions that were in revolt. The purchase was ratified in the 1819 Otis-Adams treaty that redrew the U.S. border all the way to the Pacific.

Back in Spain, an army amassed to retake the Americas instead turned on the king, forcing him to accept a liberal constitution he wanted no part of. A few months later, an uprising in southern Italy forced Ferdinand VII's uncle and namesake, King Ferdinand I of the Two Sicilies, to accept the same thing, also against his will. This return of revolution was too much for the absolutist monarchs of Europe to take. The revolt in the Two Sicilies was swiftly put down by Austria, and after three years of liberal rule, the French king restored his Bourbon cousin to power in Spain. By then, Mexico had declared itself independent, joining other states such as Chile, all of whom the reactionary kings of Europe refused to recognize.

On the fourth of July 1821, Adams gave a speech telling the Old World powers to stick to the Old World and leave the New World for the nations of the Americas—the continent wasn't open for any more colonization, or recolonization. This policy would become Adams's most famous contribution to history, even if it got named after the guy he worked for—the Monroe Doctrine.

What Adams considered his finest work, however, he had completed a few months earlier, when he delivered his report on weights and measures to Congress. Published as a book, the report has been called both a classic and a seminal work in the field of metrology; a light beach read, however, it is not. In the report, Adams tackled not only the American situation but the entire width and breadth of the subject of measures. He touched on their prehistoric origins, be it in body parts or the shells of bivalves, quoted the Old Testament for the dimensions of Noah's ark (300 cubits long by 50 cubits wide by 30 cubits high), and detailed the long legislative history of English weights and measures, the creation of the metric system, and the current state of U.S. measures before finally coming to his conclusions and recommendations.

Although few in his lifetime would ever read it, the report would go on to be cited in metric debates forever after, at home and abroad. Partly, this is on account of how it appealed to those on both sides of the issue, as Adams enthusiastically held not only the metric system, but also the English system of weights and measures to be nearly perfect. Only in a former time, however, and in part because of its kinship to classical systems.

All Roman weights and measures were, in principle, drawn from the foot, just as the French system was from the meter. One cubic foot of space equaled a *quadrantal*, which weighed eighty pounds when filled with water—or rather, the pound was set to weigh one eightieth of the quadrantal's water weight. All other weights and measures were just ratios of the cubic foot and the pound derived from it.

Adams noted that the quadrantal was the forerunner of the bushel, and that its eighth part—the *congius*—was the equivalent of the gallon, which had been defined in the thirteenth century as weighing eight pounds when filled with wine. A pound, in turn, was made up of 240 silver pence, meaning that coinage was also integrated into English weights and measures. Being both based on nature and utterly uniform, Adams found this system close to ideal.

By 1821, however, such was far from the case, and Adams found contemporary English measures—and thus American ones—to be in a deplorably corrupt condition. The villain of the tale, according to the secretary, was government. Kings who debased coinage broke the uniform perfection of the system; once a pound sterling weighed a little less than other pounds, what use did coins have as weights? Of course, any change that didn't respect the original relationships broke the system, and for this he could blame legislators in general.

The arc of this tale—that it was the bad acts of kings that sent a cherished ideal down the tubes—fit the worldview of someone who at the age of seven was drilling with the militia of revolutionaries fighting for what they believed had been theirs by right in a more perfect past. In measures as in government, Adams saw no hope for a return to the ideal, but even if such a return were possible, Adams wouldn't have wanted it. After all, the English pound, foot, and pint could never be truly universal, and universality was something Adams most ardently desired, just one reason for his mighty crush on the meter.

"A great and beautiful system" is how Adams began his ravished inquiry into France's unique measures. The meter's derivation as a portion of the great circle of the earth and the meridian survey's role in establishing its length bestowed glory upon the measure, the Academy, the people of France, and mankind in general so far as the secretary was concerned. A devotee of astronomy, Adams wished for America to be more like France in regard to great projects and state support of science, both of which he advocated.

Adams of course admired the metric system for its interrelated measures of length, capacity, and weight, and he was especially enamored of its nomenclature. The metric system had one name to one measure, making impossible something like the *ton*, which in Britain meant 2,240 pounds and in America 2,000 pounds, and when spelled "tun" referred to a measure equal to 252 gallons, which when filled with water weighed a little more than 2,100 pounds.

This didn't mean that the secretary didn't have nits to pick with the meter. He thought the French had done a poor job of integrating coinage into their system, considering the franc's nondecimal five grams of weight. And it could of course not escape a man who daily measured distance by his own stride that there was something convenient and universal about human-based standards that the metric system could never match. Adams also knew "decimation"—as division by decimals was still called—to be a double-edged sword. Certainly, he believed in the sense and simplicity of decimals, but he recognized that the English system better fit the mathematic capabilities of the average citizen. He also pointed out that the dollar's divisions had not been nearly so successful as Jefferson believed, as Spanish pieces of eight were still being passed around and calculated in terms of shillings by shopkeepers, while dimes were by and large "state secrets."

But far and away his greatest problem with the metric system—and the reason Adams couldn't recommend its adoption despite all his swooning—was the tenuousness of its very existence. He noted that the Dutch king had recently reinstituted a peculiarly localized version of the decimal metric system, but outside the United Netherlands, things looked grim for the meter.

With a reliably reactionary Bourbon king hostile to all things revolutionary, France could hardly be counted upon to long retain even the watered-down *mesures usuelles* of Napoleon. In fact, at the very time the Adams report was being delivered, the French crown was reconsidering its entire system, as were Britain and Spain. Adams's great wish was that instead of each nation deciding for itself they should come together to agree on a universal system of weights, measures, and coinage, whatever it might look like.

Such wishes aside, what did Adams advise in the end? Inaction, mostly. He had some suggestions on how to make the existing system work better, which included the federal government's sending proper modern standards to all the states and customshouses so they would be the same across the nation. But when it came to fixing even the worst aspects of the system, such as its three different gallons, Adams was against any tinkering, let alone the full decimation Madison and Jefferson advocated. It was a deflating prescription after such exuberant enthusiasm, but it squared with his belief that legislative cures were worse than metrological diseases. The one great positive of the present state of affairs was that America and Britain shared the same measures; if the United States changed its system, then the metrical rapport between the ever-expanding American republic and the great British Empire would be ruined, and how would that serve the interests of universality?

The flip side to Adams's recommendation was that Anglo-Saxon conformity depended upon Britain's not changing its measures either, something he had been assured by the British ambassador would not happen. Adams should have known from his reading of history, however, that in this he was likely to be sorely disappointed.

Imperialism

In America, the War of Independence is seen as a great victory, as it indeed was. But the reverse—that the Revolution was a great defeat for the British—is not true. For them, the colonial insurrection was a slap in the face. A discomfiting, embarrassing, and costly slap, but nothing worse than that, and in the long run, some argued, even a good thing.

While the Americans and French were having their political revolutions, the Brits were launching the Industrial Revolution. James Watt's steam engine began transforming Britain the same year as the Declaration of Independence; in the 1820s, the marriage of steam to rail was under way in English coal country, with ramifications nobody could have expected.

However much they were in the vanguard of technological innovation, the English could be conservative to the point of quirky when it came to measurement. The use of tally sticks to keep accounts at the Exchequer gave way to Arabic numeral bookkeeping only in 1724, but old tally sticks continued to be in use, with thousands still current as the 1820s began. Britain had been almost the last Western European nation to adopt the Gregorian calendar, waiting 170 years, and while the United States and France were using decimal currencies, the British had recently explored and rejected the idea. So it could come as little surprise that they would not be quick to abandon the pound and foot. The old measures were clearly no barrier to technological innovation, or the Industrial Revolution would've happened somewhere else.

Some British subjects, however, did want reform. James Watt had written in the 1770s of the desperate need for decimal measures, while others were embarrassed by the Exchequer's continued acceptance of tally sticks, which would finally be put to rest in 1826. The decade-long debate over weights and measures that officially began in 1814 with war's end spurred various radical overhaul proposals. In the end, Parliament wiped away virtually all previous laws relating to measurement—close to a millennium's worth—while at the same time attempting to preserve tradition, and created the *imperial system* of weights and measures.

Standards were narrowed to just two, both to be kept in the House of Commons. For these primary standards, Parliament retained the pound and yard fabricated in and in use since the mid-1700s. For capacity, all

previous standards were chucked, to be replaced by a new, singular measure for both wet and dry substances, derived from the weight of water.

A *pint's a pound the world around* may have been the popular saying, but
a pint of water weighed—and in our system still does weigh—about 4
percent more than a pound. The imperial reform, however, chucked this
rough equivalency in favor of reviving the Roman quadrantal. Sort of.
They set the imperial bushel to equal eighty pounds of water, which in
turn bloated its eighth part, the gallon, to ten pounds, and the pint to
1.25 pounds, giving rise to the decidedly less catchy *A pint of pure water
weighs a pound and a quarter*. It also set the imperial pint to 20 ounces,
which severed another connection between pint and pound—that both
contained 16 ounces.

The British now had one measure where there had formerly been
three, at the cost of Anglo-American uniformity. Even worse, by creating a new standard rather than choosing an existing one, there were now
four different measures called the pint being passed around the English-
speaking world.

While Adams hadn't wanted weights and measures to follow the dollar
and Webster's blue speller in forging a separate American identity, that is
what happened, although not by choice. The British imperial system
created an American system of weights and measures by dint of
abandonment.

Presidential Nadir

The Imperial Weights and Measures Act was passed by Parliament in
the summer of 1824. That fall, John Quincy Adams ran for president.
Although being the son of a former president was of questionable political capital in America (at least in those days), being secretary of state
was the surest path to the position, as every president since his father—
Jefferson, Monroe, and Madison—had used it as a launching pad.

Despite the track record, Andrew Jackson thumped Adams in the
popular vote, and beat him soundly in the Electoral College too. Jackson,

however, fell short of a majority in either. Henry Clay, Speaker of the House of Representatives and the man who placed fourth in the election, thought Jackson unqualified and threw his support behind Adams, who in turn appointed Clay to the president-making secretary of state position. Jackson was furious at what his followers would forever call the "corrupt bargain," but Old Hickory kept his cool, shaking Adams's hand with seeming cordiality at a reception held by President Monroe the day after the decisive ballot in the House.

Adams could have used some of Jackson's political canniness. In his first address to Congress, Adams didn't search out areas of agreement with his opponents the way a president elected with 30 percent of the vote should have. Instead, he presented a bold agenda that emphasized—of all things—science.

The history of governmental science in America had essentially been nonexistent. The only federally backed scientific organization was the Coast Survey, approved by Thomas Jefferson in 1807. Other than the purchase of some equipment and the measurement of a very short baseline, however, little had been accomplished, and the department was defunded. States without coasts saw the project as a waste of money, never mind how many ships were coming to grief on uncharted reefs.

In Adams's speech, the new president called for the establishment of a national university, governmental research into weights and measures, and—most important, to his mind—a national observatory. Astronomy remained the essential science, the one that birthed other disciplines; the burgeoning field of statistics was being developed by men trained as astronomers, and practical matters such as mapmaking, navigation, and meteorology depended on it utterly. And yet, not a single observatory could be found in the Americas, as opposed to upwards of 130 European "lighthouses of the sky." Adams himself had earlier offered $1,000 of his own money to get one started at Harvard, but it had as of yet come to naught.

Over the years, Adams's address has been called the most eloquent paean to government's responsibility for science ever delivered by an

American president. At the time, however, the speech was a political debacle, as Adams's own horrified cabinet had told him it would be. The savvy Clay saw it as bad politics and wondered how much of what Adams proposed was even constitutional, while others saw the new president as staking out his support of a large, centralized government, which Americans generally equated with tyranny and monarchism, a judgment further invited by comparing the republic's lack of support for the sciences unfavorably with that of foreign despots. That his intentions were pure made Adams naive at best, out of touch and detached at worst. The public wanted nothing to do with his proposals and mocked Adams's lighthouses in the sky as castles in the air.

In America to witness the early days of the second Adams administration was a figure from the president's European boyhood, Lafayette. The last surviving general of the Revolutionary War, Lafayette had returned to the country he loved for a kind of farewell reunion tour that set off the most delirious outpouring of affection the United States had ever seen. Arriving in New York to the salute of cannon fire and a crowd of fifty thousand, Lafayette spent more than a year riding a wave of nostalgia across all twenty-four states and into the nation's capital, where he would address Congress. He visited the tomb of his "father" at Mount Vernon, Thomas Jefferson at Monticello, and John Adams in Massachusetts, and he was thrown a sixty-eighth birthday party by John Quincy Adams at the White House, where he would finish off his tour with a month-long stay. For a man who loved attention, it was heaven.

Lafayette's tour was the last gasp of the Revolutionary generation in America, whose passing could be dated to the country's fiftieth anniversary, July 4, 1826, when old friends and enemies John Adams and Thomas Jefferson died within hours of each other.

Lafayette, however, was unready to exit the stage of history quite yet. France was wrapped up in its own nostalgia for revolution, more acute than America's, and had another one. Lafayette played a critical role in the July Revolution of 1830, rejecting overtures to run France himself and instead giving his imprimatur to Louis-Philippe, thereby sealing the ascent of the so-called citizen-king, a man who had spent part of the Revolution in American exile like Talleyrand and so many others. The events in France would lead to the successful Belgian insurrection against the Dutch king, which began in August, as well as the failed November uprising of the Poles against the Russians; various Italian revolts were suppressed as well. Earlier the same year, the independence of Greece from the Ottoman Empire had become a settled reality. These were all cracks in the old hegemony favored by the members of the old Holy Alliance, and stirrings of the nationalism that would come to dominate the world.

The events in Europe took place during one of the only periods in his life that Adams was not holding a political office, having been beaten too soundly in his rematch with Andrew Jackson for electoral politicking to matter. It was probably for the best. Adams stands as perhaps the only man for whom the presidency was the nadir of his political career. As his years as a diplomat and secretary of state would be marked by brilliance, so, too, would his performance in the job he held next.

Science in America

Entering the House of Representatives at age sixty-three, Adams finally found the role ideally suited to him: agitator. Adams quickly became the most important antislavery legislator in the country, despised among the Southern delegation and lionized as "Old Man Eloquent" by abolitionists. His dedication to the cause would get him a featured place in a Steven Spielberg movie—*Amistad*—but Adams would become the most powerful congressional voice on another issue as well, his predilection for which had helped ruin his presidency.

Adams's big dreams for national science had gone nowhere while he was in the White House, but in his first years in Congress the outlook improved, if only a little. In 1832, funding was restored to the Coast

Survey and it was charged with setting up an Office of Weights and Measures. This was good news for Ferdinand Hassler, an exacting and exasperating Swiss émigré who been hired to lead the Coast Survey at its inception twenty-five years earlier. Hassler would go on to be considered the father of American federal science and the man who put the United States on course to having a true set of weights and measures. Mostly, this meant creating physical standards for distribution across the country, but first Hassler had to decide exactly what those standards should be. In his so doing, the United States officially rejected Britain's imperial system, and also shed the ale gallon, leaving America with a single measure for liquids, a small victory for uniformity.

As for those imperial measures, they would never be the same after October 16, 1834. That evening, the eleventh-century Palace of Westminster—the iconic home of Parliament—burst into a spectacular conflagration. The building also housed the standard yard and pound, which were similarly destroyed. The suspected cause of the blaze was the defunct tally sticks of the Exchequer, which had been recklessly crammed into the furnaces beneath the House of Lords. "Those preposterous sticks!" Charles Dickens would later thunder, holding up the tally sticks as a symbol of the archaic traditions Britons clung to, and the fire as the consequence of such a mentality.

While the British puzzled over how to re-create the yard, the meter was receiving the greatest boost in its history. The "French metrical system," as it was almost universally called, had continued to languish for years after Adams feared for its existence. But the Revolution of 1830 was grounded in the grand ideals of the original French Revolution, which manifested itself in the return of the tricolor to France and calls for the restoration of the decimal metric system with its long, prefix-laden names.

The first nation to bring back the metric system, however, was Belgium, which had broken free from the Netherlands and was itching to establish its Frenchness and shed its Dutchness. It adopted the franc of five grams of silver first, and the rest of the definitive Jacobin-era metric system shortly after. France followed with its own restoration in 1840, shedding the Napoleonic *systeme usuelle*. Faraway Greece, itself full of revolutionary fervor, also declared itself metric, although in practice the coins and measures of its hated former Turkish overlords continued to be used.

In addition to its newfound political popularity, the meter was becoming well established in European science. Americans going abroad in increasing frequency in the 1830s to apprentice at Continental observatories became familiar with it, as did those working at home for the Coast Survey, which thanks to Hassler had been using the meter since his measure of its first baseline back in the 1810s, meaning that the most important science project in America had been metric from the start.

Meanwhile, Adams's greatest dream was coming within reach. In 1835, the American people inherited a large sum of money that its English donor, the eccentric James Smithson, had earmarked for a scientific "establishment" to be based in Washington. Made chair of the House committee charged with making use of the bequest, Adams pounced on the opportunity to create a national observatory. Bill after bill, however, came to naught.

But even without federal backing, the 1830s saw the opening of America's first permanent observatories. They were modest affairs when compared with European ones; instead of being based in capital cities and funded by absolute monarchs, they tended to be at colleges and driven by the support of a single individual. The first serious homes for astronomy came in 1839, when Harvard secured financing for its observatory, and 1842, when Congress finally authorized the money for what would become the Naval Observatory, which marked a turning point in the thus far meager tale of federal American science.

The following year, the Coast Survey was transformed when Hassler died and his position was assumed by Alexander Dallas Bache, the grandson of Benjamin Franklin. Bache had been one of the first

Americans to study in Europe (where he became an ardent believer in the metric system) and was well on his way to becoming the central figure in U.S. science. More efficient and a better manager than Hassler, Bache would radically speed up coastal charting and soon finish the job of creating sets of measurement standards to be distributed nationally.

Bache would also be on the Board of Regents that established the Smithsonian Institution in 1846, and he persuaded the other great American scientist of the day, Joseph Henry, to accept its secretaryship. Far more than just a museum, the Smithsonian under Henry would conduct scientific research such as collecting meteorological data from across the country to create the first national weather maps.

"Lighthouses of the sky" kept popping up across the land, spreading to the South and Midwest. They included the first purely professional observatory in Cincinnati on Mount Ida, which was renamed Mount Adams in honor of the seventy-seven-year-old former president, who presided over the laying of its cornerstone. Though slowing down, the old amateur astronomer delighted in the nights he spent at the Naval Observatory, where, not far from his home in Washington, he could gaze up at the stars through a state-of-the-art telescope that was the property of the American people.

On February 21, 1848, Adams had a stroke on the floor of the House. His last word entered into the congressional record was "No!" It was spoken against a motion thanking U.S. generals for their services in the Mexican War, an engagement Adams had fiercely opposed.

By that time, the young year had already seen two earth-altering events. The first was the discovery of gold in Sutter's Mill, California, and the second was the outbreak of the most dramatic season of European revolutions yet. News of neither, however, had reached Washington by the time Adams died, two days after his collapse. For not much longer would news travel so slowly, however. The world was about to change, practically all at once.

or One Half

THE INTERNATIONALISTS

A week before John Quincy Adams fell stricken on its floor, the House had been presented with a bill supporting the kind of worthwhile yet seemingly quixotic cause that Old Man Eloquent had been fond of. This one concerned cultural exchange, a field Adams had been introduced to by Alexandre Vattemare.

Vattemare was like no man before or since. A ventriloquist, he was one of the first international celebrities. The Frenchman had been meant for a career in medicine but was kicked out of school for throwing voices on cadavers one too many times. Nevertheless, he was pressed into medical service at the end of the Napoleonic Wars, and he found himself stranded in Berlin at conflict's end. To make a living he took to the stage, performing his "vocal illusionism" in hundreds of cities across Central and Eastern Europe. Fame and success on a grand scale came in the early 1820s when Vattemare moved to London, where his self-scripted one-man shows with multiple characters won him acclaim alongside the greatest Shakespearean actors of the day. Wealthy, Vattemare returned to France to begin his second life.

While on tour, Vattemare had obsessively visited local museums and libraries, which he discovered to be heavily stocked with duplicate pieces. He created a system whereby institutions could exchange such redundancies across national borders. Failing to gain traction in Europe,

Vattemare took the recommendation of his friend Lafayette to concentrate his cultural exchange efforts between France and the United States.

Arriving in America to great fanfare in 1839, Vattemare visited John Quincy Adams numerous times. The crotchety old legislator complained that the Frenchman "bedaubed me with flattery till I sickened," but Adams was taken with his cause—as well as his ventriloqual feats—and introduced into Congress a memorial on cultural exchange penned by Vattemare. (In the nineteenth century, a "memorial" was a statement of facts by which lawmakers were petitioned.) Although a bill passed in support of his program, actual exchanges languished until Vattemare returned in 1847, this time bearing the *Description de l'Égypte*, the masterwork created by Napoleon's army of savants.

During his extended stay, Vattemare organized for complete sets of weights and measures to be swapped between the United States and France. At the time, Alexander Dallas Bache was delivering his final sets of American standards, ensuring at last that a California pound weighed the same as a pound in Maine. These state-of-the-art measures and balances were a testament to Bache's skills as a scientist and an administrator, but in his report to Congress, the Coast Survey chief said he would gladly see them become immediately obsolete, believing that the nation should adopt the metric system.

As Bache noted, his plea followed one made a half year earlier to Congress by the secretary of commerce, Robert J. Walker, who desired that all nations be on the same decimal system of coinage, weights, and measures. The man behind the free-trade-inspired Walker Tariff, the commerce secretary represented a different kind of Democrat. They were called New or Progressive Democrats, but the catchall term for their movement was Young America. The Young Americans had one of their own as president, James Polk, and believed it was the manifest destiny of the United States to stretch out across the continent spreading democracy and progress. Theirs was an exuberant nationalism that looked with positive eagerness at the revolutions then exploding across Europe.

The very name Young America had been lifted from European nationalist movements. The original was Young Italy, created by the greatest liberal nationalist of all, Giuseppe Mazzini. Mazzini's organization had inspired Young Germany, Young Ireland, Young Hungary, and other Young Europe movements, most of them created by men hungering for liberation from foreign despotism and nations of their own, and potential builders of American-style democracies.

As ever, the headline was the revolution occurring in France, which swept out the citizen-king and ushered in the Second Republic. This revolutionary spring had started in the Two Sicilies, spread up the Italian peninsula, and exploded in the states of the former Holy Roman Empire, where the Frankfurt Parliament formed with the aim of creating a single German nation. Most inspiring to Americans were the Hungarians with their dashing and eloquent leader Louis Kossuth, who attempted to expel the hated Austrian Hapsburgs, the most active of the Holy Alliance in suppressing revolution over the years.

Political progress had in many ways been spawned by technological progress, of which the 1840s had seen an unprecedented explosion. Transportation by steam (in shipping as much as rail) had reached critical mass, as had the telegraph in America. The greatest revolution in everyday communication, however, had come not from Morse's wires but from the postage stamp, which was first introduced in 1840. The Penny Black took a letter anywhere in Britain for 1d—a little more than two U.S. cents at the time—a drastic reduction in the price of mail, previously charged for by the mile. By 1848, correspondence in the United Kingdom had more than tripled, and the cheaper postage had allowed for the mass circulation of journals and newspapers. Such freedom of communication was anathema to the Holy Alliance, who could only watch as the new French Republic introduced the stamp among its very first acts.

The postage stamp was the brainchild of a liberal reformer named Rowland Hill and was linked to the greatest liberal cause of all—free trade—which also had its breakthrough in 1840s Britain, with Parliament's repeal of the Corn Laws ("corn" being the British term for wheat). This reform also crossed borders, with the Walker Tariff one example. Ideas could be exchanged much more easily than Vattemare's objects. Alexis de Tocqueville had visited the United States and gone home to write *Democracy in America*, which examined the system of government he felt was coming to France. Traveling the other way was the codification movement in law. The year 1848 saw the debut of the Field Code, the brainchild of David Dudley Field, a Young America Democrat from New York who sought to wipe away the archaism of customary law in his own country as the Napoleonic Code had done elsewhere.

All of the advances of the decade—the stamp, the telegraph, regular rail travel, transatlantic steamship passage—had created a growing belief in the limitless virtue of Progress, the invariably capitalized idea that had become the narrative of the era. Some believed that a new world was at hand, one in which national borders would be erased and an everlasting peace achieved. In the autumn of 1848, people looking to bring about such eventualities assembled in Brussels at the First International Peace Congress. Before the meeting was even finished, its organizers started planning a blockbuster sequel for the following year, a Peace Congress that would take place on the world's grandest stage—Paris—and have for its chair the outstanding literary personality of the day.

Proposals for Peace

The man who strode onto the podium in the large Paris concert hall in August 1849 was familiar to all in attendance. They had read his poems and books and knew his face, even if his forehead was somewhat less enormous than caricaturists made it out to be. His hair was long, befitting the author who had brought Romantic poetry to France, even if those locks had begun to look awkward when pasted against his increasingly jowly face.

Victor Hugo had been the most famous poet and writer in France for a generation, having burst onto the literary scene as a teen. Hugo was the proto-celebrity activist, fighting for the same causes as actors and pop stars do today. His first novel was a condemnation of the death penalty, while his international bestseller *The Hunchback of Notre Dame* had helped spark the architectural preservation movement in France, the old cathedral being a widely reviled wreck at the time. What brought him to head the Peace Congress was his grandest cause, that of the establishment of a United States of Europe, which he believed would bring peace on earth, and the case for which he made in his opening speech.

"A day will come when there will be no battlefields, but markets opening to commerce and minds opening to ideas," Hugo said. "A day will come when the bullets and bombs are replaced by votes, by universal suffrage, by the venerable arbitration of a great supreme senate which will be to Europe what Parliament is to England, the Diet to Germany, and the Legislative Assembly to France. A day will come when a cannon will be a museum-piece, as instruments of torture are today. And we will be amazed to think that these things once existed!"

Hugo's speech would become a signature of the peace movement, and later earn him a place as a forefather of the European Union. At the time, however, it mostly earned him ridicule. The Romantic poet did himself no favors with lines like "We must love each other!" as solutions for ending war.

Charges of naive utopianism and being just plain silly had dogged the peace movement since its birth in New England and Great Britain in the days of Waterloo, when years of Napoleonic conflict spurred groups of Quakers to band together. The "plain people" still formed the heart of the movement, which contributed to its image problem, with the bonnets of "Quakeresses" attending the Peace Congress providing plenty of second looks and chuckles on the fashionable streets of Paris.

But the idea of a United States of Europe, especially when linked as a step to a single world government, sparked fears of a world monarchy and Napoleon-like tyranny. Rather than a universal state, U.S. delegates tended to back a Congress of Nations. A principal promoter of this idea was the organizing force behind the Peace Congress, the so-called Learned Blacksmith, Elihu Burritt. At the heady peak of the 1848 Revolutions, the Yankee Burritt had hoped to recruit twenty U.S. congressmen and senators to participate at the Paris event, and imagined a Congress of Nations as being just four or five years off. The chief goal of such an organization would be the institution of a standing court to settle disputes between countries and to administer a code of international law. Attendees could argue over which was better, but few outside the Salle de Sainte Cécile believed any of it had a prayer of coming to pass. Some inside the auditorium knew it too, most particularly Richard Cobden, whose support gave the Peace Congress its greatest claim to legitimacy.

Cobden was at midcentury one of the world's leading figures. A textile manufacturer turned radical member of Parliament, Cobden possessed a kind, handsome face parenthetically enclosed by blond mutton chops. Cobden's years-long struggle to overturn British grain tariffs had made his name synonymous with free trade, and his Anti–Corn Law League had become the organizational model for liberal reform movements of all stripes, the Peace Congress included. He had also campaigned for the Uniform Penny Post, which provided Cobden his first taste of success.

The tariffs of the Corn Laws—or Bread Taxes, as Cobden called them—increased the cost of living for workers while lining the pockets of wealthy landowners. Their 1846 repeal caused Britain's international trade to boom (just as the stamp had done for the mail), to the great benefit of farmers in the American Midwest, an especially appealing side benefit for Cobden, who believed the United States to be the model of peace and prosperity for the world.

For Cobden and other liberals, free trade—be it in commerce or correspondence—was not just about saving the common man money, but about peace, as the more nations depended on one another, the slower they would be to go to war. He called it "the moral equivalent of gravity" for drawing people together. Cobden brought this practical approach to the Peace Congress, where he pushed for narrow, politically achievable resolutions.

The international congress was, at midcentury, a phenomenon still in its early stages. Congresses were being called to discuss women's rights, meteorology, sanitation, prison reform—nearly anything—and were creating a framework of international cooperation where none had existed. The Peace Congress was a typical example, drawing delegates from local peace societies across Europe and America to a foreign capital where they would vote on resolutions that set the agenda of the movement. These resolutions were spread by reports published on the proceedings, as well as by delegates returning home to get the word out through speeches, literature, petitions, memorials, posters, and propaganda envelopes, which were something like bumper stickers for your letters.

Two of the resolutions that Cobden pushed for were international arbitration and disarmament, which would become the main objectives of the peace movement for the rest of the century and beyond. Cobden argued for a system of arbitration because he knew that any kind of permanent international tribunal would be unacceptable to practically every lawmaker and ruler in the world. Other pipe dreams such as Hippolyte Peut's suggestion for a universal language Cobden also opposed, wondering how you could ever decide which to choose. Cobden did approve, however, of another proposal by Peut that called for a universal system of weights, measures, currencies, and postal rates. Not only would they diminish borders, unite people, and promote freedom, they also seemed achievable. Others agreed, and Peut's resolution passed handily.

The business of the three-day Peace Congress at an end, delegates met for a Saturday night soirée hosted by Foreign Minister Alexis de

Tocqueville that lasted until midnight. The following week, on the first of September, a delegation headed by Victor Hugo went to present the congress's resolutions to the first-ever president of France, Louis-Napoleon, the nephew of the man whose thirst for battle had triggered the peace movement in the first place.

For many, the rise of the new Napoleon was the symbol of how very wrong things had gone with the Revolutions of 1848. The Peace Congress took place at the end of a summer that had been darkened by cholera and violence in Paris, and the spectacle of Louis Kossuth and Giuseppe Garibaldi—the heroic would-be liberators of Hungary and Italy—on the run to their eventual exiles in Britain and America. In this they followed the "Forty-Eighters" of Germany, who had fled Frankfurt upon the forced dissolution of the pan-German parliament. All across Europe, constitutions granted and tricolor flags flown by frightened monarchs were shredded as the Holy Alliance—though shaken—managed once again to quell revolution.

Only the one in France continued, with a revived Assembly and President Napoleon, whose resounding electoral victory had less to do with his own record than name recognition. When Hugo handed Louis-Napoleon the resolutions of the Peace Congress, the jury was still out on him, but the prince-president gave Hugo reason to hope. Louis-Napoleon said that he agreed with the principles of the congress, and he would prove as good as his word on the subject of universal measurement. The president had earlier assured Hugo that he wanted not to be another Napoleon Bonaparte but a second George Washington—music to the poet's ears.

Despite the dark clouds, there was still much hope on the liberal horizon. In fact, what promised to be the grandest celebration of hope and Progress ever conceived was in the offing, a Great Exhibition to be held in London in 1851. And it was there that the most avid receiver of Peut's Peace Congress proposal, a man named James Yates, would see an opportunity for making universal measures a reality.

The Probabilities of Measurement

That the universal adoption of the metric system could be the most immediately achievable resolution of the Peace Congress speaks volumes about its phoenixlike rise. In a single decade, the meter had gone from all but invisible to the leading system of measurement in Western Europe. It happened especially in lands that had been under direct Napoleonic rule and where metric standards persisted in some (often bureaucratic) form.

Such a place was Piedmont, the Turin-based kingdom that was the most developed of the Italian states. In the mid-1840s, King Vittorio Emanuele decided to follow his larger neighbor and main trade partner France in reverting to the decimal metric measures, issuing a decree that would make them the law of the land in 1850. Various Rhineland states of the pan-German free-trade zone called the Zollverein ("Toll Union") either had never entirely done away with the metric system or now chose to readopt it, including the Kingdom of Bavaria, the Grand Duchy of Baden, and the Free City of Frankfurt, while newly united Switzerland reintroduced the franc in 1850. Spain resolved to adopt the metric system, although actual use of it was a long way off. Faraway Chile, the most powerful of the South American republics, similarly stated their intention to go metric and abandoned the Spanish eight-real dollar in favor of a decimal currency based on the franc.

Americans crossed the sea in unprecedented numbers to see London's Great Exhibition in 1851. Those who decided to further hop the English Channel for a Continental vacation would pleasantly discover that from Bruges to Paris and the Swiss Alps to the Italian Riviera they wouldn't have to go to the costly and annoying expense of changing and rechanging money. It mattered not whether the coins were called francs or liras, or whether they sported the monarchical heads of Vittorio Emanuele or the Belgian king Leopold or the idealized symbols of the new Swiss Confederation or the République française—the money was essentially the same. Even greater conveniences greeted importers and exporters, who dealt in metric-based measures over a similarly large area.

Such obvious benefits in travel and commerce bred evangelists. The decimal metric system had become a part of the grand catchall Progress and benefited from the perception of being new. Not that the meter's acceptance was smooth or fast. It always met popular ridicule, and legacy measures persisted everywhere. Wines were still shipped in the same old *boisseaux*, clothmakers used the ancient *aune*, and road signs were marked off in *miglia*. But for the first time, decimal weights and measures were not only the law in public markets but actually being used there. The moment was not lost on Frenchmen like Hippolyte Peut and Alexandre Vattemare, or Americans living in Paris, and especially not the envious Englishman James Yates.

A member of the Royal Society, Yates had been so impressed by Peut's advocacy of universal measures that he contacted the French canal engineer with an idea for how to make a metric splash in his home country. Yates wanted to organize an exhibit of the French metrical system at the Great Exhibition in London, which was at its heart an international industrial fair. The exhibit would not only demonstrate the universal utility and efficiency of decimal measures but also expose the yards, ells, *braccia*, *punds*, and *tummes* that countries used to tout their mechanical accomplishments as the preposterous sticks they really were. To lend Yates the sets of standards he needed, Peut turned to the Paris-based Conservatory of Arts and Crafts.

The brainchild of the Abbé Grégoire, the Conservatory was a museum and teaching organization dedicated to sciences and trades housed in a former monastery suppressed during the Revolution. It contained the world's greatest collection of weights and measures and had recently been given responsibility for creating and calibrating metric standards for foreign countries, making it for all intents and purposes the headquarters of the metric system. The Conservatory was headed by Jean-Thiébault Silbermann, a zealous proponent of metric standards who was only too happy to arrange a display for London.

The Great Exhibition itself wound up being a success beyond even the most grandiose dreams of its patron, Prince Albert, and enthusiastic commissioners like Richard Cobden. Attendance exceeded an unthinkable six million visitors, who arrived from across the world to ooh and ah over the enormous Koh-i-Noor diamond, the most recent treasure the British had extracted from India, to puzzle at inventions like the Alarm Clock Bed, which woke its inhabitants by flipping them into a tub of water, and to complain about the food, dull even by English standards.

But more than any one exhibit, the main impact of the Great Exhibition was the event itself. It was the first World's Fair, done on a scale to make every previous exposition seem puny and provincial by comparison, and housed in the world's first glass-clad building, the Crystal Palace, itself the greatest marvel of the event. Exhibits were dedicated not only to industry and inventions but to different aspects of world history, culture, and art, their overarching theme not only Progress but the peace and prosperity it would bring. The Great Exhibition would forever be seen as a turning point and moment of creation, including the birth of internationalism, a term that came into the language in 1851.

The celebration of mankind as a whole caused nationalist prejudices to be set aside, especially in Great Britain, which became awash in internationalist sentiment. A new Peace Congress running concurrently with the Crystal Palace expo was not laughed at, instead receiving respectful and even positive notice in the press. For Victorian peaceniks, it was a dream come true, with talk of a new era of peace being heard everywhere. Peace was *in*.

While the machinations of Yates, Peut, and Silbermann did succeed in kicking off the British metric movement, an even bigger consequence of the Great Exhibition as far as the meter and franc were concerned was the fair's impact on Adolphe Quetelet, a man interested in the universal measure of everything.

Quetelet was the astronomer royal of Belgium, but more important, he was the world's leading statistician, his name practically synonymous with the profession. A stint at the Paris Observatory put Quetelet in the same building with such statistical pioneers as Laplace, who inspired him to follow in their footsteps. The Belgian did so with an eye toward social issues, applying probability theory to behaviors such as crime; he also created the concept of the average man, for which he developed the body mass index (BMI), still known as the Quetelet scale.

At the Great Exhibition, Quetelet met with the heads of state statistical bureaus from around Europe and came up with the idea of creating an International Statistical Congress. In some ways, this would be like other congresses in that it would be promoted by local interest groups, in this case statistical societies whose memberships included anyone interested in political economy. (The term *statistics* was derived from "statist" for a reason.) But Quetelet's congress would also include government representatives, creating a forum for cooperation between nations in a day when such a concept was unheard of.

The first International Statistical Congress met in Quetelet's home city of Brussels in 1853. Quetelet's greatest hope for the organization was that it would come up with a universal language of statistics, to render them more easily comparable and thus of greater utility. A shared system of decimal measures would greatly aid this goal, and one of the congress's resolutions was to ask that all national statistical reports include at least secondary sets of metric figures. But this was only a toe in the water; the statistical congresses would soon become the central forum in the international metric push.

In the meantime, more countries were jumping on the metric bandwagon, including Portugal and the future Colombia (although neither would actually give up its old measures for decades). Even the German Zollverein flirted with the metric system, establishing the use of a half-kilogram Zollpfund at its customshouses.

American interest was also expanding. In 1852, the Conservatory of Arts and Crafts sent the U.S. government a full array of metric standards, the same as they would a nation intending to convert. This completed the swap set in motion by Alexandre Vattemare, who celebrated by sending Congress a petition signed by 146 U.S. citizens living in Paris who begged their representatives to adopt the "French Metrical Decimal System," which they claimed to be the best yet invented. Vattemare also forwarded a memorial written by one such American expat, the journalist William Wilberforce Mann. Mann believed that the foot and pound were bound for the same fate as knee breeches, the optical telegraph, and stagecoaches, and he argued that the United States must adopt the meter at once, no matter the costs. A similarly urgent memorial came to the Senate from the American Statistical and Geographical Society, a new organization headed by the powerful Young America Democrat George Bancroft.

All the various threads—the meter's momentum and the work of Peut, Yates, Vattemare, Mann, and the Conservatory, as well as the interest of Cobden, Quetelet, and others—would come together at the second Statistical Congress, where the desire for a universal set of measures would coalesce into an international movement. It would happen, fittingly, in Paris.

The International Association

The Exposition Universelle of 1855 was the French response to London's Great Exhibition. This second World's Fair took place in a Paris where entire medieval city blocks were falling en masse according to the designs of the new city planner, Baron Haussmann, a side consequence of the coup that erased any optimism that the new Napoleon might be a democratic champion of the people. In short order, Louis-Napoleon had declared the Second Empire to be in effect and himself Napoleon III, rendering the Second Republic no more. Personifications of Liberty were soon replaced on French coins and stamps by portraits of the new emperor.

The change in regime was bad news for peace congresses. France went from being a source of inspiration and membership to a place where freedom of speech was kaput. A voice like Victor Hugo's could now be heard only in exile. What finished off the pacifist movement, however, was the autumn 1853 outbreak of the Crimean War. In England, to be for peace meant to be against the war, an unpatriotic stance at best. Peace was out, war was in.

If the 1855 Exposition was Napoleon III's international coming-out party, the same could be said for the metric movement. In the meter's native land, authorities did what they could to promote the system. An exhibit by the Conservatory of Arts and Crafts compared the kilogram, meter, and franc to the weights, measures, and coinages of the rest of the world, with an eye to highlighting how diverse and backward the latter were, while the imperial commissioners of the Expo asked all exhibitors to use metric measures in their materials. The request was little heeded, but more than two hundred jurors and commissioners signed a petition complaining that the old measures rendered their jobs impossible and stated their wish for a universal system of decimal measurement, a desire echoed in a strong resolution by the second International Statistical Congress, which was held alongside the Expo.

Behind both the petition and the resolution was a pro-metric circular getting passed around to officials of the Expo and delegates of the congress. This document was the handiwork of James Yates and signed by, among others, Richard Cobden and six of his fellow members of the House of Commons. Similarly, Alexandre Vattemare (then serving as one of the U.S. commissioners for the Exposition) was circulating copies of W. W. Mann's pro-metric tract, bound together with a supporting paper by Conservatory director Silbermann. These hardcore metric advocates took things one step further, forming the all too literally named International Association for Obtaining a Uniform Decimal System of Measures, Weights, and Coins.

The first official gatherings of the International Association were held by invitation of Napoleon III in the Emperor's Room of the Palais de l'Industrie, the exhibition center erected for the Exposition on the Champs Élysées. The inaugural speech was given by the Baron James de Rothschild, founder of the French branch of his family's banking empire and first president of the new group.

Rothschild was just one of an immensely influential and powerful cast of individuals who joined the International Association. Its British chapter boasted the likes of Erasmus Darwin and Charles Babbage, as well as the ever-reliable Cobden, while its American roster would soon include George Bancroft, Alexander Dallas Bache, and Matthew Fontaine Maury, director of the Naval Observatory. Lesser known but more emblematic members were staunch internationalists like the Belgian Auguste Visschers, chair of the first peace and statistical congresses, and Leone Levi, an Italian Jewish lawyer and statistician who had emigrated to England, become a Presbyterian, and authored some of the earliest works on the metric system, international currency, and a universal legal code.

At its founding, what the International Association did not have was complete agreement over whether a single universal system of measurement was even possible. Visschers, for instance, believed the chances for such an outcome were slim, imagining the best one could hope for was a winnowing down to three, four, or five systems. Backing his notion was the fact that the most powerful nation in Scandinavia—Sweden—was planning to decimalize its customary measures the following year, pointedly choosing to modernize without going metric. This followed a long tradition of the Swedes being opposed to the French measures, seeing their native ones as vital for the preservation of national independence.

The idea of decimalizing customary measures was also gaining traction in Great Britain. Of course, this had long seemed a commonsense solution to many, including Madison and Jefferson in the 1810s. John Quincy Adams, on the other hand, had recognized that such a turn would doom the prospects for a universal system, a sentiment shared by

Hippolyte Peut, who believed it better for England to keep what she had rather than go through the wrenching process of decimalization. Once decimalized, Brits and Swedes would be loath to change again. From that perspective, the opportunity for a universal system of measurement was a window that was already closing.

One member vehemently opposed to any outcome save a universal system was the American George Sumner, a die-hard metricalist who was anxious for the United States to drop not only the yard for the meter, but the dollar for the franc. Another expatriate living in Paris, Sumner had helped his fellow New Englander Elihu Burritt organize the Peace Congress and had been in on the hatching of the International Association but had recently returned home after fourteen years abroad, largely out of disgust with Napoleon III. Back in America, Sumner would lecture his countrymen on the virtues of the French Revolution, at times to fierce approbation. For men like Sumner, the connection between the metric system and French-style republicanism was not trivial, representing as it did the best of the Enlightenment values of savants such as Condorcet, while Laplace's 1799 scientific conference was held up as the forerunner of the international congresses then being used to reshape how the world functioned.

The country George Sumner returned to didn't want to listen to such things at the moment, as it was in the process of withdrawing from the international stage it had so shortly before been thinking it would dominate. The United States was falling into the black hole of sectionalism, a crisis precipitated by no one so much as another member of the International Association, George's older brother, Senator Charles Sumner.

Unions

Charles Sumner had gone to Europe before his brother, among the first of the brilliant young Americans to go abroad in search of furthering their education. A Harvard-trained lawyer, Sumner had gone to study the Code

Civil in Paris with the aim of bringing its principles to American law, but he got sidetracked by other reforms that captured his ever-fervent mind. His pacifist 1845 Fourth of July lecture *The True Grandeur of Nations* brought him wide acclaim, and made him a pillar of the peace movement. Elihu Burritt had dearly wanted him to come to the 1849 congress, but Sumner was otherwise occupied radicalizing American politics.

Abolitionism had become his leading cause, for which he had taken up the mantle of most-despised Yankee from John Quincy Adams, who in his final years had become a mentor to Sumner. With the ex-president's son Charles Francis Adams, Sumner led the Massachusetts delegation of the new Free-Soil Party, which nominated Adams for vice president and Sumner for Congress. Both failed to get elected in 1848, but three years later Sumner became a senator under the Free-Soil banner.

While new to the Senate, Sumner championed universal postage rates and the Statistical and Geographical Society's pro-metric memorial, but before long he was delivering extremist speeches on abolition and slavery. Sumner's focus became increasingly single-minded, culminating in a purposefully provocative address he delivered during the Bleeding Kansas crisis of 1856 that nearly ended his life and upended the equilibrium of the nation.

The speech took as one of its targets Sumner's fellow senator Andrew Butler and Butler's "harlot, slavery." Sumner's oratory enraged the plantation-owning politician's cousin, South Carolina congressman Preston Brooks, who decided to meet verbal violence with the physical kind and attacked Sumner two days later in the Senate chamber with his gutta-percha cane. Sumner, trapped under a desk bolted to the floor, was beaten into bloody unconsciousness. Fined $300, Brooks was hailed as a hero in the South, while Sumner became a Northern martyr.

With his brother George overseeing his care, Charles's health waxed and waned, and he would be unable to fully return to work for three years, during which time he spent two stints convalescing in Europe.

Back in the Old World for the first time since his youth, Sumner felt invigorated. He renewed old acquaintances and hobnobbed with a virtual who's who of liberal Europeans, including Alexis de Tocqueville and Alexandre Vattemare. Vattemare took Sumner around Paris numerous times and introduced him at the Conservatory of Arts and Crafts, where Director Silbermann chatted up the senator over universal measures. There were multiple meetings with Cobden, as admiring of America as ever, and also Cobden's close associate Michel Chevalier, a world-respected liberal economist whose work was widely read on both sides of the Channel, a rarity.

As with so many other important internationalists, Chevalier was a Peace Congress alumnus and member of the International Association, whose annual meeting he would chair in 1859. It was held that year in England, where Chevalier arrived with an almost outlandish proposition—that he and Richard Cobden engineer a free trade agreement between France and Britain.

Though skeptical of Chevalier's idea, Cobden was willing to take a flier on it. British alarms had been sounding ever since another Napoleon had taken control of the French state, fears that had grown acute that summer of 1859 with Napoleon III's military success at the epic Battle of Solferino. The decisive moment in Piedmont's victory over Austria for supremacy in northern Italy, Solferino also led to the French annexation of Savoy and Nice, the price extracted by Napoleon III for his military support. Concern over the emperor's next move provoked invasion hysteria, never far from the surface in the British Isles. The prospect of a costly race to build arms on both sides of the Channel—not to mention their potential use—vexed Cobden immensely.

At the secret meeting Chevalier arranged between Cobden and Napoleon III, the French emperor listened attentively to the Englishman outline the benefits of free trade, responding that it was just the kind of thing he'd like to do for his people but that he doubted it could happen. "We make revolutions in France," he said, "not reforms."

Still, Napoleon III was determined to liberalize his regime and, putting his weight behind the agreement, negotiated it successfully, to the shock of both nations. The 1860 Cobden-Chevalier Treaty proved a watershed that sparked dozens upon dozens of similar agreements across Europe, creating the Continent's greatest free-trade era prior to the advent of the European Community.

It seemed, in fact, as if a dam had broken, with several liberal dreams being realized all at once. The long-desired unification of Italy was achieved, with the new nation adopting in full the metric measures of its new sovereign, the king of Piedmont. In South America, most states had by the early 1860s adopted the meter in principle, and even if none had yet put it into practice, the new Ecuadorean *franco* and Peruvian *sol* joined Chile's peso as franc-based currencies.

In 1862, a sequel to the Great Exhibition brought a new swell of universalist goodwill to England. On a path paved by the free-trade treaty and the determined efforts of the International Association, a Select Committee of the House of Commons convened to investigate metric adoption. In testimony, the postage stamp creator Rowland Hill heartily endorsed the meter, a position in which he was joined by a cavalcade of International Association members that included Chevalier, James Yates, and Leone Levi, as well as individuals from the scientific, educational, and commercial communities. Cobden was on the committee, and he proclaimed that he had never seen such unanimity of opinion in all his years in Parliament. His fellow members of the House of Commons were convinced, and the bill to adopt the meter passed on July 1, 1863, by a wide margin. It was too late in the session for the bill to move to Lords, but few among those cheering doubted its success.

The idea that the British Empire would be going metric in three short years transformed international opinion. With length measures based on the English inch, Russia had stated it would follow the British lead, and few doubted America would as well. What was more, a postal congress slated for the following year looked to deliver a universal stamp.

One problem darkened the spirits of internationalists. America, the continent-straddling land of one language, one measure, one coin, one stamp, and free trade among a brotherhood of states—the model nation of uniformity and peace—had gone to war with itself.

⁹⁄₁₆

A UNIVERSAL COIN

O N MARCH 3, 1863, the first year of the Thirty-Seventh Congress finally came to a close. Amid empty seats vacated by their Southern counterparts, overworked congressmen and senators finished up their business inside a half-finished, domeless Capitol building that was symbolic of a Union both broken and never truly forged in the first place.

In the waning hours of the workday, a new bill was taken up by the Senate. Though not entirely read, it passed quickly, a story repeated in the House. The bill was signed by President Lincoln and became law that same day. This wasn't how Congress usually worked, not only for such speed and lack of attention but for what the bill contained. It established a National Academy of Sciences, for America an unprecedented, European-style organization of the sort John Quincy Adams had so dearly desired. Fifty men were preordained to be members for life, a group chosen by the tight circle of individuals who had set the ball rolling on the Academy, at the center of which was Alexander Dallas Bache, the long-serving head of the Coast Survey.

The way the National Academy bill got slipped through Congress was widely condemned as underhanded, but the real outrage came over the undemocratic and elitist way in which its members had been selected. Many disliked the Bache clique, both in the ever-expanding Coast Survey (its mission having been broadened by the war) and the Lazzaroni, as Bache's group of academic pals called themselves. Smithsonian head Joseph Henry, a close friend of Bache, blasted the "ruse" by which the Academy bill had become law, but he swallowed his disgust and pledged to do his utmost in making the new organization useful to the war effort.

The first assignment of the National Academy was the establishing of a committee to look into universal weights, measures, and coinage. Two more committees quickly followed that were both dedicated exclusively to coinage, a vital issue during any war, and an obvious concern of Treasury Secretary Salmon Chase. Chase had already strongly recommended to Congress the adoption of universal decimal measures, but on currency he favored the unilateral setting of the American dollar to the British pound.

A dollar-pound union had been a live issue for years, with a delegate having been sent by the United States to Britain in 1857 to explore a single currency for the two nations. A unified Anglo-Saxon monetary system would have been the culminating moment of a friendship that had been building since the end of the War of 1812 and reached a fever pitch in 1858 when the first messages on the transatlantic cable were exchanged between Queen Victoria and President Buchanan. The event was hailed as being on a par with Columbus's voyage, but the cable broke after just three weeks.

Warm U.S. feelings toward its former motherland pretty much ceased once the English began to dally with the Confederacy. Charles Francis Adams, the third in his line to be ambassador to Great Britain, managed feelings in London, while his political comrade Charles Sumner kept Lincoln apprised of the at times shocking level of anti-American war sentiment in the British government, thanks to inside information he was being fed by Richard Cobden.

During the Civil War, the Union wanted little more from its foreign policy than to keep the European powers out of the conflict. Napoleon III, ever ready to pounce on the strife of others, flouted the Monroe Doctrine by intervening in Mexico and installing a Hapsburg "emperor" there. Early on, some European liberals favored recognizing the

Confederacy, on the grounds that it had established itself as a separate nation; others favored the Union but held deep reservations about its failure to abolish slavery, the reason why Garibaldi, the hero of both Italy and Uruguay, turned down an offer from Lincoln to become a general on a third continent. Garibaldi would lead European liberals in cheering Lincoln's Emancipation Proclamation when it went into effect at the beginning of 1863, the year the Civil War turned in favor of the Union. It was also the year that the U.S. government would begin reengaging the wider world, sending delegates to a pair of international congresses. These delegates would bring the metric movement home to America, and one of them had a plan for a universal currency that just might work.

25 to 5 to 1

When Samuel Ruggles attended the 1863 Berlin Statistical Congress, he became the first official U.S. delegate sent to such an event, beginning a career that would see him be America's man on the international congress scene for a decade. A New Yorker, Ruggles had served on the Erie Canal Commission, created Gramercy Park and Lexington Avenue, and helped transform Columbia College into a major university. Energetic and ebullient, idealistic but practical, Ruggles was the perfect man to enhance the United States's image abroad.

The statisticians gathered in the Prussian royal city of Berlin were lobbying hard for the metric system, and *only* the metric system—there was no more talk of a vague "universal uniform decimal system" or having multiple systems coexist. What's more, the resolutions of the 1863 congress asked that, once adopted by a particular country, the "Metrical System" be made compulsory in the shortest possible order. It was a rip-the-Band-Aid approach that had been written into the recent House of Commons metric bill, which set a three-year deadline.

Even more notable at the new Statistical Congress was how hard opinion had swung toward a universal currency. While the desire for one had long been expressed, the likelihood of its happening had been doubted. Although the franc was integral to the metric idea, many internationalist advocates didn't stump for it out of fear that the greater metric movement might get derailed with the extra complications that came with issues of national currency, even though the franc had been spreading faster than the meter for years.

There were also those who, like John Quincy Adams, didn't care for the franc on account of its weight having been set to match the old *livre* rather than a perfectly decimal ten grams of silver. Making things worse was the *napoleon*, the world's most widely minted gold coin. The napoleon was worth twenty francs, but because it was gold and the franc defined in silver, its weight was based on the exchange rate between the two metals, set by France at 15.5:1. This meant that a gold franc weighed .322581 grams, making it about as metric as the .304801-meter foot. Purists like the free-trader Michel Chevalier believed this to be a "breach" that compromised the metric system's integrity. Chevalier and those of like mind wanted weight to equal value; ideally, names like "franc" would be redundant, and coins could simply be stamped ONE GRAM GOLD.

Few thought such a coin possible. Pragmatists aimed for a reduction in the number of currencies via unions, a trend already under way. The German states of the Zollverein and Austria entered a currency pact in 1857 that fixed the values of their individual coins against one another and created the Vereinsthaler, while unified Italy had seen all coinage save the lira disappear. There was also the potential Anglo-Saxon currency union, brought up again at the Berlin congress by a British delegate. Ruggles favored the idea, but wanted to take it a step further: He believed a universal coin could be achieved *through* the union of the dollar and the pound, if both were in turn set to the franc.

A pound was worth very close to five dollars, which in turn was valued very close to twenty-five francs. A universal currency based on the three could have as its standard a coin worth £1-$5-25F. Similar ideas had been floated before but usually focused on unifying currencies based on silver, whereas Ruggles had his sights squarely set on gold, and only gold. The difference was crucial; there were countless silver coins in circulation around the world, but just three important gold ones: the British sovereign (£1), American eagle ($10), and French Napoleon (20F). This was true for the simple reason that Britain, France, and America were the only sizable nations maintaining their own gold standard. The rest of the world (or nearly) was on a silver standard.

A UNIVERSAL GOLD COIN

Coin	Denomination	Weight (grams)	Value (francs)
Napoleon	20F	6.452	20
Half-Eagle	$5	8.359	25.91
Sovereign	£1	8.136	25.2
Universal	1 unit	8.065	25

This is the plan proposed by Samuel Ruggles at the 1863 International Statistical Congress in Berlin, which became the centerpiece of discussions to unify the moneys of the world during the 1867 International Monetary Conference in Paris. The weight of the sovereign is given as if it were 9/10ths fine (90 percent gold, 10 percent alloy); the coin weighed slightly less because the British used the duodecimal Crown standard of 11/12ths fine. That all coinage be ".900 fine" was another resolution of the 1863 Statistical Congress.

Of course, France and America were also on a silver standard. Officially, they were both bimetallic; in practice, the United States was not. The Forty-Niners of California had dug up so much of the shiny yellow stuff that its price tanked, eventually hitting the point where silver dollars were worth more than gold ones in terms of the market price of metal they contained. Speculators bought up silver dollars to sell abroad where they got melted down and recoined into rix dollars and thalers, forcing the United States to devalue and then all but stop minting silver coins.

The absurdity of such a situation was not only economic but philosophical. To have two standards of value—gold and silver—was as absurd as having different pints for blueberries and beer. The burning question was which metal should be the standard.

Opinion went back and forth. In silver's favor was the fact that the majority of the world's currencies were set to it. Gold, on the other hand, had the sheen of Progress. Partly this owed to Britain's seemingly limitless prosperity since becoming the first country to go on a gold standard almost half a century earlier, and partly because silver was so highly prized in Asia, the place that epitomized backwardness. The first was only slightly more logical than the second, but one indisputable fact was that gold was a lot more convenient.

A silver dollar was a fair-sized disc of metal—four times the size of our quarter—whereas a gold dollar was a good bit smaller than our current dime. If you bought a horse for $180 from someone insisting on payment in coin, you could either lug a ten-plus-pound sack (avoirdupois) of 180 silver dollars to the stable, or slip into your vest pocket a ten-ounce purse of nine double eagles.

In terms of creating a universal currency, what made gold even more attractive was the relative clean slate it provided. Rather than having to accommodate the currencies and concerns of so many silver-based countries, the number of coins that needed coordinating was three, which was in a way only two. With the war, the Union had suspended payment in specie and gone on the paper greenback, a major reason why some Americans were so anxious for a universal coin—at the moment they had no real currency at all. To have that universal coin be gold was an extra bonus, considering that the United States held vast deposits of it.

At Berlin, Ruggles's plan met with enthusiasm. However, the statisticians realized that the topic's scope was beyond them, a matter for economists and diplomats. In the end, a universal currency was outlined at the Statistical Congress, with the stated need for the governments of the world to come together to put it into action—a meeting that was still a few years off. But 1863 did see a near agreement made on a different kind of universal currency: the stamp.

Pleasure at a Double Purpose

To send a letter internationally was as much a mess in 1863 as it had been before the advent of the postage stamp, which generally worked for mail sent only within the country that issued it. Letters posted internationally racked up charges everywhere they went. An envelope sent from America to Europe would be charged the U.S. domestic rate, the sea rate, the rate of each country it passed through, and the one in which it was received, as well as various "registration fees" and other such nonsense. Its journey might go any of four, five, or six different routes, each of them subject to a different tangled web of treaties, all of which had been negotiated separately according to the weight, measures, and currencies of the countries involved. Not only the complexity but the expense was horrendous—a single letter could cost over a dollar per half ounce when sent to Europe, whereas the rate from Chicago to San Francisco was three cents.

How this mess could be fixed was obvious: a universal stamp that could take a letter anywhere. This was why international postage had been grouped with other universal measures at the Paris Peace Congress, and why its organizer, Elihu Burritt, had dedicated himself to the cause once the peace movement hit the skids. Charles Sumner was also taken by the idea and championed the "ocean penny post" (as the idea of a cheap, one-rate transatlantic stamp was known) throughout his years in the Senate.

The 1863 Paris Postal Congress set down a list of thirty-one rules of how a universal stamp should work. It resolved that there should be one international rate based on weight, which for a standard letter should be set to 15 grams, the nearest round metric equivalent to the penny post's half ounce. Although not resolved at the congress, the rate was also discussed, and this, too, would be metric, with a 15-gram letter costing 25 centimes, a little less than five American cents. In fact, it was explicitly stated that *only* the metric system should be used in the post, showing that delegates considered it to already be the lingua franca of measurement.

The International Postal Congress had been called by U.S. Postmaster General Montgomery Blair and was attended by his former second in command, Rep. John Kasson. The Alexandrian solution conceived of at the congress—that the tangled knot of international postal rates could be solved by a single metric chop—deeply impressed Kasson, who had just been elected to Congress and quickly became its most bullish member on both the meter and the franc. The following year, he formed the House Committee on Weights, Measures, and Coinage, which would deliver its first report in 1866, when legislators would be assembling under the newly completed dome of the Capitol.

Chosen on the same ballot as Lincoln's reelection and the first to sit after Appomattox, the Thirty-Ninth Congress was filled with Radical Republicans bent on imposing harsh conditions for the reentry of Southern states into the Union. These "Jacobins," as they were called by their enemies, had been at odds with the more forgiving Lincoln over the issue while he lived, and become only more hardened since his assassination. The session's signature piece of legislation was the Civil Rights Act, which granted full rights of citizenship to the former slaves, a position outraging many in the north, let alone the former Confederacy; nevertheless, it passed over President Andrew Johnson's veto in April 1866. The following month, John Kasson's committee submitted its findings, along with a bill to make use of the metric system legal in the United States.

The Kasson report followed closely on the heels of another extremely pro-metric report delivered by the National Academy of Sciences. This earlier report would serve as the basis of the metric movement in America for years to come, particularly in the practical recommendations it made to Congress, which included the distribution of metric standards to states and customshouses as well as the exclusive use of metric weights and sizes in the post and with new coins. In addition to relying heavily on the Academy's findings, Kasson appended to his bill petitions and testimonials—one had "the mayor, judges, and citizens of Baltimore praying for the adoption of the metric system"—as well as the report of the 1862 select committee of Parliament, no matter that the British metric movement itself had gone off the rails.

The passage of the metric adoption bill had caught Britain off guard—it was as if no one had believed the talk was serious. The most powerful newspaper in the world, the London *Times*, was incredulous at the thought Britain would be giving up its yard and pound in three short years. "What are France, the Zollverein, and Portugal to us?" the paper wrote shortly before the metric bill passed the House of Commons. "They are accustomed to revolutions, earthquakes, and wars." Opposition mounted and the bill never made it to the House of Lords, but its supporters soldiered on, well aware that nowhere had the meter been adopted easily.

In America, prospects looked brighter, with Kasson's Metric Bill passing the House and moving on to the Senate, where it had for a sponsor the Radical Republican Charles Sumner. The simmering July heat of Washington lessened none of Sumner's trademark grandiloquence, and the senator proved he could gush over the French metrical system with the best of them. After its acceptance, the system would be "among the choicest possessions of an advanced civilization," right up there with Arabic numerals, and Sumner believed that its being based on the measure of the earth was a "transcendental" idea. Its adoption, he claimed, would also help lead to peace.

"There is something captivating in the idea of one system of weights and measures," Sumner said, "which shall be common to all the civilized world; so that, at least in this particular, the confusion of Babel may be overcome. Kindred to this is the idea of one system of money. And both of these ideas are, perhaps, the forerunners of that grander idea of one language for all the civilized world."

When the metric bill passed in July, it was actually the second piece of metric legislation that Kasson had shepherded through Congress that year. The first had eliminated the half-dime in favor of a new five-cent piece, the coin we call the *nickel*.

The half-dime had been inconveniently sized from the start—no bigger than a pinkie nail—possessing as it did half the amount of silver of the already small dime. The new five-cent piece was considerably heftier, nearly four times so. As with all American coins, it didn't have a portrait on it, such a thing being considered the antidemocratic mark of despots like Napoleon III. (Not until the 1909 Lincoln cent would the likeness of an individual appear on a U.S. coin.)

But the new five-cent piece was far from just another American coin. Made of a copper-nickel alloy as hard as any ever used in currency, the coin positively gleamed with modernity. With a laurel-draped shield on its front and a great numeral 5 ringed by bursting stars on its back, the new coin looked as different from the shillings and worn-out pieces of eight still circulating as a steam locomotive from a horse-drawn carriage. Less obvious but of greater import was that each coin measured exactly two centimeters in diameter and weighed five grams.

Among the first to get their hands on the new five-cent piece was Samuel Ruggles, who took a sampling to Northampton, Massachusetts, where the National Academy of Sciences was holding its summer meeting in the comfortable surroundings of the Berkshire foothills. In an air of triumph, the exultant Ruggles held up the freshly minted coins for those gathered to see, an audience of the foremost scientists of the nation that included Joseph Henry, Louis Agassiz, Benjamin Peirce, Ogden Rood, and Wolcott Gibbs. Ruggles threw one down on the table and declared, "There, gentlemen, is your money. It is coined at your request. It will serve you at pleasure a double purpose—to weigh your letter or to pay your postage."

The new five-cent piece had achieved the holy grail of coinage, where value equals weight—a cent for every gram. What's more, three of them would balance the weight of one standard letter, and—if Ruggles's universal currency plan took hold—each of them would be worth 25 centimes, the future cost of sending that letter anywhere in the world. Beyond a weight, the nickel could serve as a ruler, with five of them making a decameter, the metric "hand" of approximately four inches. More than a mere coin, it was the metric system put into the pockets of every American citizen, linked to the promise of universal communication.

It was but a first step. Kasson had also been among those asking Congress to authorize a representative to go to the forthcoming Exposition Universelle in Paris, where an international monetary conference was being held with the aim of creating a world currency. That representative would be Samuel Ruggles, and it was his plan—or something very much like it—that Kasson and other internationalists hoped would win the day.

The Greatest Expo of All

The second Exposition Universelle of Paris ran from April Fool's Day to Halloween 1867 on the Champ de Mars, the site of the future Eiffel Tower. The latest world's fair was a showcase for the new Paris as physically remade by Haussmann, and people who knew it from before Napoleon III's now sixteen-year reign could hardly believe the transformation. Out of a twisting forest of medieval alleys, Paris had emerged as something positively geometric, with wide boulevards radiating out from grand open spaces.

As a world's fair, the new Expo trended away from nourishing exhibition and more toward carnival. It being French, the food was better than at any London fair, but more important, it was just plain bigger than anything that had come before, with 52,000 exhibitors, nearly seven million visitors, and forty-two nations participating. As had become de rigueur with such events, there occurred encounters pregnant with future import, such as Jules Verne getting ideas from a submarine display and an exhibition of Japanese woodblock prints that would influence the direction of Western art.

A high-minded take on the goings-on was put forth by the introduction to the *Paris-guide de l'exposition universelle*, which declared a world's fair not to be about a comparison of products, as it ostensibly was, but of utopias. The Exposition was a harbinger of the future, the introduction's author maintained, of a twentieth century in which an extraordinary nation would emerge, both rich and peaceful, that would make wars on the Continent obsolete. "This nation will have Paris for a capital, and will not be called France; it will be called Europe." The writer looked even further into the future, when this great nation would be transfigured still more and include the entire world, and be called simply Humanity.

This internationalist vision was written by a man still in exile from France, Victor Hugo. No more the long-haired Romantic of his Peace Congress days, Hugo now embodied the role of the fiery, white-bearded elder, and his words rang out more as prophecy than the poetic fantasies they seemed in 1849.

The ongoing consolidations of Italy and Germany were wiping away the medieval jumble of states at Europe's heart the same as Haussmann's boulevards had cleared the congestion of Paris. The defeat of disunionists in America had led to a more integrated, continent-spanning nation, proof that the model worked. The long-cherished dream of a United States of Europe seemed attainable, as did a future world with "unity of language, unity of money, unity of measure," as Hugo put it, and as surely as Paris would be the capital of this place, the French language, the franc, and the meter would be its standards. On this point, at least, Hugo and Napoleon III could agree.

To showcase the weights, measures, coins, and stamps of the world in all their perversity and diversity, imperial commissioners had erected a two-story domed pavilion dedicated to their display in the central courtyard of the main exhibition building. Its importance could hardly be missed at an expo that was playing host to two congresses on measurement, the first—and more important—being the International Monetary Conference.

Attended by twenty nations, the conference was in essence a full-court-press sales job for the franc and the organization administering its rules, the Latin Monetary Union. Formed in 1865 by France, Belgium, Switzerland, and Italy, the franc union was attracting enormous interest, with nations including Spain, Greece, Romania, and Austria flocking to apply.

In early 1867, an invitation to join had been extended to the United States and was welcomed in some quarters. Kasson and Ruggles both approved of the idea, but its most important American supporter would prove to be John Sherman, chair of the Senate's powerful Finance Committee.

Arriving in Paris ahead of Sherman, Ruggles found that his coin unification plan was being embraced by the most powerful economist in the French government, Felix de Parieu. Before he could push it at the International Monetary Conference, however, Parieu needed to know that the U.S. would adopt the franc. Assurance to this effect was provided by Sherman, who wrote that he was positive that the Senate would do whatever it took to make a universal currency a reality.

Thus backed by Sherman, Ruggles became one of the key international players in the days leading up to the monetary conference, his exuberance helping to sway opinion within a French government that was divided over switching to the gold standard. Those Ruggles met with included Napoleon III, whose foreign escapades had lately been turning out badly. The emperor had mishandled French bargaining power in the Austro-Prussian war and had been embarrassed by Otto von Bismarck in an attempted after-the-fact negotiation, while his North American dalliance had turned into a debacle, with the imminent execution of his puppet, the Emperor Maximilian. The franc becoming the basis of a universal currency could provide a much-needed boost to France's international reputation.

"Franc," however, was hardly an appropriate name for the new world money. Some at the conference wanted to recycle a great name of the past like *ducat* or *zecchino*, while others preferred the high-minded *cosmos* or *unité*; the choice of the architect of the Latin Union and leader of the conference, Parieu, was the *europe*. Although it made little sense if the United States joined, this name showed where the heart of the project—and Parieu—lay, as did his desire for calling the monetary club the Western European Union. It was almost as if the EU were forming 125 years ahead of schedule, with a free-trade zone stretching from Portugal to Austria and Sweden and discussion of a European parliament.

While long trumpeted as a means to end currency speculation and make trade more efficient, a universal currency was being increasingly pushed as a path to a more united world. The £1-$5-25F gold coin or something similar was wanted by most of the delegates attending the conference, while those holding out for something purer, like Chevalier, were dismissed as metric zealots.

As for the conference itself, there were two principal questions. Would gold or silver be the universal standard of value? And would countries be willing to adopt a 25-franc universal coin as the basis of their currency?

The greatest impact of the conference came in how that first question was answered, with a sudden and near total commitment to gold. Coming in, only two out of the twenty participating nations were on an official gold standard; by the end, only two out of twenty were not at least resolved to go gold. In short, the international gold standard was born in 1867 Paris.

Before switching to a gold standard, however, most countries wanted to see how things played out with the second question. Ideally, the move to gold and a universal coin would happen in conjunction. Holding up such a move—just as Napoleon said they always would—were the British and the Germans.

If Great Britain had wanted a single coin for the world, it would have happened. The British, however, stood to lose the most from such a thing. The pound sterling already *was* a universal currency, used in shipping accounts the world over. Having the strongest of many currencies was highly profitable for the United Kingdom, and a pillar upon which

London's place as the center of international finance was built. To change, Britain would not only have to voluntarily give up its perch on top of the heap, but remint all its coins so they could circulate freely with those produced by other nations. And what Englishman wanted the portrait of Napoleon III in his pocket?

The other kingdom clogging up the works was Prussia. Its situation was thornier than Britain's, as it was trying to position itself as the crown under which Germany would unite. The latest step in that process was the North German Confederation, a collection of twenty-two states that formed as a result of the previous year's Austro-Prussian War, a kind of German civil war in which Prussia and its northern, mostly Protestant allies defeated Austria and its southern, largely Catholic brethren. The battle for German hegemony won, Prussia began to court defeated southern states like Bavaria and Baden, which in turn feared unification as nothing more than a Prussian takeover. Largely situated in the Upper Rhine Valley, some of these states had been within French borders for two decades in Napoleonic times, when they had thrived using French laws and measures. They saw the franc and meter as bulwarks for independence, and a way to preserve their historic ties to France and Western Europe. Although the Prussians preferred that the pan-German currency be the silver *Vereinsthaler* or "Union dollar" (which was little more than the old Prussian *thaler* with a more politic name), they couldn't be opposed to a franc-based world coin for fear of scaring off Bavaria and the rest.

In all countries, though, there were people who opposed the idea of a single universal coin, just as there were those against the meter. Some thought it natural that there be German, Anglo-Saxon, Latin, and Scandinavian sets of measures and moneys, the same as there were families of languages. The concept of a "Latin America" had begun to emerge, promoted in part by Frenchmen looking to extend their influence. Not everyone was impressed, however, with the "Latin Monetary Union" being in origin a derisive British term. For the franc (or whatever the new international coin would be called) to become truly universal, it would have to break out of its Latin pigeonhole. The same went for the metric system.

In Germany, arguments for and against metric adoption mirrored those for a universal coin; Prussians preferred their own standards, Bavarians and their neighbors the French ones. What was different, however, was that here science was as much in play as politics. Prussia's foot was based on the steel-and-sapphire *Urmaass* created in the mid-1830s by the German astronomer Friedrich Bessel, while Britain's new yard had been produced in 1845, the result of a decade-long project to replace the one destroyed in the Parliament fire.

The meter, on the other hand, had been made in a rush in 1799, with platinum now understood to be impure, using outdated methods of fabrication. There was also no international organization administering the metric system as there was with the franc, let alone a global corps of measurement inspectors to match the British Empire's. Its standards were a French possession, and the most important task that fell to its keepers—creating and calibrating standards for other nations—was carried out not by a modern laboratory but by the Conservatory of Arts and Crafts, a teaching institution and museum.

These issues came into focus at the meeting of the International Committee of Weights and Measures that ran concurrently with the monetary conference. On July 2, an international nine-man delegation from this committee went to the Hôtel des Archives, where they were allowed to examine the original meter, nestled in a long mahogany case lined with red velvet and bearing a brass plaque inscribed METER CONFORMS TO THE LAW OF 18 GERMINAL, YEAR III, PRESENTED 4 MESSIDOR, YEAR VII.

The field trip had been organized by Moritz von Jacobi of the Russian Academy of Sciences. Jacobi had been born in Prussia, a not-unusual origin in the world of Russian science, where the working language was

German. Having for years lobbied for Russia to adopt the French system, Jacobi was an enthusiastic proponent of the meter, just not the one sitting in the Archives, let alone the one kept in the Conservatory, where Jacobi's delegation headed next.

The meter of the Conservatory was the one used to make copies, so any discrepancy between it and the meter of the Archives was being habitually perpetuated. Even more problematic was that varying methods and materials had been used over the years to make duplicates, creating a chaotic situation in which the meters of various nations were not in agreement. While the measurement pavilion at the heart of the Champ de Mars sought to expose the world's nonmetric measures as preposterous sticks, it had become clear to many scientists that the meter had itself become a problematic one.

Europe Divided and United

As the 1867 Exposition Universelle entered its final month, the International Conference on Geodesy got under way in Berlin. It was part of the largest cooperative, international scientific undertaking in history, with echoes of the original meter survey and Laplace's 1799 conference that poured over its results. The project was, in essence, a joint survey of Central Europe. At the time, national maps fit together like the pieces of different jigsaw puzzles, having been created using varying prime meridians and measures. Unlike in the U.S. with the Coast Survey, the meter was little used by geodesists in Europe, who continued to employ the old *toise*; that the meter was not yet the measuring stick of those who measured the earth had long been one of the great knocks against it.

At the conference, a strong consensus emerged that the transnational survey go forward using the meter, but only if the meter were fixed. And not only in terms of a new physical standard; delegates wanted to create an organization of international professionals to oversee construction of a new meter and administer it. The resolutions they passed had extra bite because the Central European survey was a prestige project of Prussia and had the weight of that increasingly powerful nation behind it. When the North German Confederation announced it would go metric as of January 1, 1872, it was fair to wonder whether it would be using the French version of the system or something different.

The turn of events did not sit entirely well in certain quarters of French science, where some believed the meter and kilogram to be both sacred relics and their perpetual property. France had neglected to send any representatives to the geodesy summits in Berlin, and there was a general anxiety over the ascendancy of Prussian science and the feeling of having been eclipsed. There was also fear that a new meter would be established without French involvement, based on the new measure of the earth that the Central European survey would provide.

Fortunately, good sense won out over wounded national pride, and France got behind efforts to create a new meter, with Napoleon III sending out invitations for a new international metric conference to take place in the fall of 1870. Most could see that German adoption of the metric system was a great victory for a French creation, making it no longer a purely Latin instrument. There was also reason to believe that the Prussian-dominated North German Confederation would next adopt the franc.

Austria, Sweden, Spain, Greece, and a gaggle of smaller states had all either joined or were in the process of joining the Latin Union, or had begun to strike coins based on its standards. New applications were coming from as far afield as Brazil. Senator Sherman had put forth legislation to adopt the franc in America, while in Britain the Chancellor of the Exchequer in the new Liberal government supported modifying the sovereign to conform to international standards. The idea of one coin that could be used as easily in Frankfurt as San Francisco was more than fantasy—it was on the cusp of happening.

As was an international meter. The United States took the 1870 Paris metric conference seriously, sending as a delegate its foremost scientist,

Joseph Henry, who was now president of the National Academy of Sciences, the organization he had once so opposed. The first meeting of the conference took place at the Conservatory of Arts and Crafts, where Henry greeted his fellow scientists from other nations. Not attending, however, were the Prussians. Not their scientists, at any rate. Their soldiers, on the other hand, would be in Paris all too soon.

Few had thought this would be the case when France declared war on Prussia that summer. At the time, there was a deeply ingrained, Continent-wide fear of the French, who hadn't been beaten on the battlefield since the original Napoleon's day, and then only through a Herculean effort and a Russian winter. The German-speaking world was better known for art than for war, with the vital exception of autocratic and militaristic Prussia, which had on its side the brilliant General Helmuth von Moltke and the dark genius of Bismarck. In truth, France was badly overmatched. A scant three weeks after the war had begun, it was over for Napoleon III, who was humiliatingly defeated and taken prisoner on the battlefield at Sedan.

This was when things got interesting. Liberal European opinion had been heavily for the Germans—the emerging young nation versus the despotic empire. With the removal of Napoleon III, however, and the re-reestablishment of the French Republic, international perception turned, in some ways forever. Victor Hugo came home after eighteen years of exile to witness the siege of Paris, where he found his fellow citizens surviving on rodents. (He noted in his diary that rat pâté was said to be quite good.) The suffering of Parisians seemed all for nought when the new French government capitulated and the king of Prussia was crowned emperor of a united Germany at Versailles. Rage at the surrender led to the uprising known as the Paris Commune, a fourth French Revolution that compressed the worst excesses of the previous three into ten weeks, with atrocities and villains to go around on all sides.

For its troubles, the new German Empire charged France an

Thomas Jefferson (left) was devoted to the quantification of virtually everything, and wanted to implement a decimalized system of weights and measures in the United States. The only part of his vision that came to fruition, however, was the dollar, which won out over an earlier coinage plan conceived by Gouverneur Morris (right), a political rival who would succeed him as ambassador to France.

The 1792 *disme* and half-*disme* were among the initial coins produced by the U.S. Mint. The *s* in *disme* is silent; the spelling *dime* first appeared on coins in 1837.

Clockwise from top left: The Marquis de Condorcet was a pioneering mathematician and the savant in chief of France, being head of its Royal Academy of Sciences; Talleyrand was called "shit in a silk stocking" by Napoleon for his conniving ways, but was a liberal politician who introduced Condorcet's measurement system to the Assembly; Christiaan Huygens was the inventor of the pendulum clock, the pivotal measurement device in more ways than one.

Perhaps the greatest chemist ever, Antoine Lavoisier was a true believer in decimals and a driving force behind the metric system.

"France's Newton," Pierre-Simon Laplace, did more than anyone to try and make the metric system universal in the Napoleonic empire.

The presidency was the only job John Quincy Adams wasn't good at. His *Report upon Weights and Measures* is a classic in the field of metrology, and the most influential document on the subject ever written by an American.

The grandson of Benjamin Franklin, Alexander Dallas Bache stood at the nexus of American science in the mid-nineteenth century, turning the U.S. Coast Survey into a world-renowned scientific organization.

The other giant of early American science, Joseph Henry made his greatest impact as the original secretary of the Smithsonian Institution.

The internationalists of the mid-nineteenth century believed that the path to a more peaceful and prosperous world lay in part through a universal system of measurement and money. Important figures in the movement included (top) Elihu Burritt, the driving force behind the International Peace Congresses; (bottom) Charles Sumner, who lobbied in the U.S. Senate for the metric system, cheap international postage, and a universal currency; (opposite, clockwise from top left) ventriloquist Alexandre Vattemare, whose mission was to bring about cultural exchange between France and America; Richard Cobden, the icon of free-trading liberals and pacifists; Adolphe Quetelet, who transformed the field of statistics; and author Victor Hugo, political activist and champion of a United States of Europe.

New Yorker Samuel Ruggles devised a plan for a universal currency that came close to being adopted, and helped pave the way for the international gold standard.

The shield nickel was not only America's first coin to be sized in millimeters and grams, but was believed by idealists to be an important step toward the United States adopting the metric system.

What could have been: The French government gave Ruggles four specimens of a dollar-franc gold coin to take home from the 1867 Universal Expo.

That history repeats itself as farce was an observation made by Marx in comparing the original Napoleon's coup with that of his nephew's. Being underestimated by his opponents, however, helped Napoleon III score many successes during the Second Empire, a period in which both the metric system and the franc expanded like wildfire.

Columbia president Frederick Barnard (above) was a pivotal figure in the history of higher education. He was also an important astronomer and the founder of the American Metrological Society, an organization that led the charges to make America metric and bring Standard Time to the world. One important member of the AMS was Barnard disciple Melvil Dewey (opposite top), who not only created a decimal system for categorizing books but vigorously fought for the metric system and simplified spelling. Opposed to Barnard and Dewey was the pioneering astronomer Charles Piazzi Smyth (opposite bottom), who believed that God had encoded the ideal measuring system in the Great Pyramid.

Cleveland Abbe produced the first U.S. weather forecasts, and methodically saw through the adoption of Standard Time.

Often credited with the creation of Standard Time, Sandford Fleming campaigned vigorously against the A.M./P.M. division of the clock and was an early supporter of Moses Cotsworth's thirteen-month calendar.

Opposite: Daylight saving time was first passed in the United States during the Great War. Reviled by farmers, it was quickly canceled in its aftermath.

Moses Cotsworth spent half a century fighting for a universal thirteen-month calendar. His 1902 manifesto, *The Rational Almanac*, sits at his knee in the photograph.

Heiress Elisabeth Achelis championed the World Calendar. Whereas Cotsworth would've made every month like February—twenty-eight days long—Achelis wanted to fit February into a pattern where every thirty-one-day month was followed by two thirty-day ones.

The propaganda from religious groups against fixed calendars reached fever pitch with a publication showing "Blank Day" planes dropping bombs on churches.

The crowning moment of the American anti-metrication movement was the Foot Ball in lower Manhattan's Battery Marine Terminal. Pictured at the event are organizer Seaver Leslie (right) and fellow anti-metric activist Tom Wolfe (left).

The conversion of the nation's road signage to kilometers was the most visible step toward metrication planned by the federal government. Public outrage scuttled the $100 million project, however, turning it instead into a major defeat for metric advocates.

indemnity of five billion francs. Crucially, this allowed Germany to amass the bullion reserves it needed to convert to a gold-based currency, which it decided would be the mark, not the franc. The creation of the mark was a crushing blow to those who had worked so hard for a universal money. By introducing a new competitor to the existing gold currencies, a world coin would never again be so easy as bridging the small gap between one pound, five dollars, and twenty-five francs.

It was a step the Germans took not unthinkingly but with relish. There were those who couldn't stomach Frenchmen like Hugo talking about universal capitals, measures, and languages when in reality all they wanted was for the rest of the world to be more like Paris. The brotherhood of mankind was great for the French so long as they didn't have to change at all.

By going gold, the Germans helped to force the issue for everyone, leading to a game of musical chairs in which no one wanted to be left holding the bag of silver. America stuck to the dollar, adopted a gold-only standard, and finally got ready to pay off the paper-money greenbacks. Sweden, Norway, and Denmark joined the scramble to go gold as well. Wanting no part of the German mark and permanently cut off from the franc zone, they formed the Scandinavian Monetary Union, chucking their various silver currencies for a single gold crown, creating yet another new gold coin.

All those old thalers and rix dollars had to go somewhere, and demonetized silver flooded the market. The price of silver tanked, making a dual-metal standard unsustainable. Amid massive speculation, France was forced to go gold, even though its new conservative government didn't want to, and brought the Latin Union along with it.

The stakes had been raised so high for what a universal coinage could mean—a united Europe, peace everlasting—and the object so close to being realized that it was easy to miss just how much had in fact been accomplished. The international gold standard grew out of the quest for one money for the world, and even if the same coin couldn't be used from Frankfurt to Frisco, the gold standard did achieve many of the same benefits that a universal currency would have.

There was now, indisputably, one standard of value, which meant that exchange rates could be permanently fixed. Traveler's checks, a concept just about to find wide use, could be printed with their value in multiple currencies, a great benefit to the American summer tourist, who—if he hadn't been back to Europe since the time of the Crystal Palace—would notice just how many fewer currencies there were floating around. The unification of Italy alone had erased 282 coins from production. In Germany, he wouldn't have to remember that there were 360 pfennigs to the thaler and 240 to the gulden because there were a hundred of them to every mark. With the notable exception of the pound sterling, European currencies had gone decimal virtually everywhere.

A quarter century earlier, such a situation was as good as most internationalists had dared dream. Even better was the fact that continental Europe had all but unified behind the metric system, as it became the sole legal set of measures in the new German Empire, with Switzerland and Austria soon to follow.

The meter has been called the first European integration, which is on some level true. In 1872, however, European was all it was. No matter how many Latin American countries had pledged to adopt it, they still used the old Spanish customary measures, just as America and Britain used the old English ones and the Ottoman, Chinese, and Persian empires their own. This gave the United States the opportunity to be the nation that pioneered the meter's use outside Europe, the goal of the zealous reformers who formed a new organization called the American Metrological Society.

$10\!\!\!/_{16}$

or Five Eighths

THE BATTLE OF THE STANDARDS

LOCATED AROUND the corner from the still-under-construction St. Patrick's Cathedral, whose thirty-three-story neo-Gothic towers dominated the area, Columbia University stood in what today is called midtown, but what in 1870 was the far northern environs of New York City. At the time, Columbia was at the forefront of the movement to expand colleges into universities made up of different schools, dedicated for instance to law and scientific research.

Another movement Columbia would play a leading role in was the metric one. The university was early on connected via its energetic board member, Samuel Ruggles, and was the first school to present a memorial on behalf of the metric bill to Congress in 1866. After that bill passed, Rep. John Kasson turned to the University Convocation of New York—the organization that oversaw Columbia and other schools—for help in rallying support for teaching the meter to the "rising generation." In response to Kasson's request, the University Convocation convened a panel of educators to assess the metric issue, and they chose Charles Davies to chair it.

A professor emeritus of mathematics at Columbia, Davies had in the past written glowingly of the metric system as near "perfect." Professor Davies, however, held some unexpected views on its potential use in America. These were revealed in both his report to the University

Convocation and his subsequent 1871 book, *The Metric System*, which delivered a withering dismissal of its title subject.

Many of Davies's complaints were similar to those of Frenchmen who had undergone the system's original introduction seventy-five years earlier, especially when it came to decimal fractions. As the author of a popular series of mathematics textbooks, it wasn't that Davies didn't believe that decimals were superior for calculation—he did—but Davies's concern was that decimal math was not the language of every-day calculation, and he worried that people would lose all instinct to judge size and value in the event of a changeover. A metricalist like Charles Sumner allowed that the older generation would likely never understand the new system, but Davies thought old habits would linger far longer, and as evidence he pointed to the dollar. Pieces of eight continued to circulate, and there were still Americans calculating in shillings and pence. In France, the *decime* had never taken off, with people still preferring to reckon by the *sou*, the Carolingian name for a twentieth of a *livre* that had been replaced by the five-centime coin. The old divisions didn't die easily anywhere.

Also distressing to Davies was the idea of abandoning "short, sharp Saxon words" like *foot*, *peck*, and *ton* for the multisyllabic pseudoclassical metric terms. Back in the 1850s, there was plenty of support for calling the meter a *yard* in the same way the franc got translated into *lira*, *drachma*, or *peseta*. Practical experience, however, had found that the using of old names for new measures only caused confusion in countries that converted, so opinion had turned. In the Senate, Sumner suggested naturalizing metric terminology into English with contractions like "cenmet" and "kilmet," while the oldest supporter of the metric system in America, W. W. Mann, believed that its nomenclature needed to be overhauled completely and wrote a book as to how that also came out in 1871.

Although various educators took issue with Davies's writings, his positions positively incensed Frederick Barnard, the president of

Columbia. In the 1840s and '50s, Barnard had been in the vanguard of American science, having established the first permanent observatory south of the Mason-Dixon Line at the University of Alabama. At the time the Civil War broke out, Barnard was president of the University of Mississippi, a pointless position once the entire student body had more or less enlisted in the Confederate army. Barnard went to Richmond to plead with President Jefferson Davis for a pass to cross the lines, on the grounds that his family was all on the other side and that the war had cost him his occupation. "Oh! I will find you occupation enough," said Davis. "You are the very man that I want at this time."

A Yankee who found the Confederacy repugnant, Barnard declined Davis's job offer and remained stuck in Norfolk until May 1862, when Union forces took the town. In short order, he set to work for his friend Alexander Dallas Bache at the Coast Survey, became a charter member of the National Academy of Sciences, and was hired as president of Columbia, where he would become one of the key figures in the history of American education.

At Columbia, Barnard came up against the same conservative board members Ruggles had been fighting. Although his long white beard and kindly visage would remind students of Moses, they were a mask for a rebellious spirit that rarely shied from a fight, especially if one of his causes was being attacked. His strategy was to overwhelm; famously long-winded, Barnard was not one to pause and listen to others. This was at least partly due to the near-complete loss of hearing that had begun to afflict him as a young man, although many on the Columbia board with whom he clashed were sure that Barnard used his deafness as weapon, purposefully not recognizing objections, to their exasperation.

Davies's antimetric report and book triggered Barnard's battle instinct, for several reasons. First was the substance of his arguments, which Barnard vehemently disagreed with; then there was the fact that Davies had written on the subject at the behest of New York State educators, and instead of expressing their views, Barnard felt Davies had betrayed them. Making it all the more galling was Davies's affiliation with Columbia.

In 1872, Barnard published his own book called *The Metric System*. It was a full rebuttal of the previous year's tome, which Barnard would call a "shrewd scheme" by Davies "to scatter his crude views at the expense of the State of New York." He accused the old math teacher of padding his slim volume with an enormous appendix—in a 327-page book, Davies's contribution ended on page 54. (John Quincy Adams's report and an essay by the legendary Sir John Herschel filled out the rest.) In addition to assailing Davies's various arguments in particular and the absurdity of customary measures in general, Barnard sang the praises of the metric system, relating its history and what a boon it had been for Europe.

Barnard was exceptionally well qualified to discuss such matters, serving on the weights, measures, and currency committee of the National Academy and having been the lone American scientist on Jacobi's 1867 international commission that investigated the state of the original meter and kilogram.

His book functioned as a kind of metric manifesto and call to arms. Barnard proclaimed the meter to be inevitable, and that America faced a stark choice between embracing Progress or being left behind. It was no longer just European, he noted, with Brazil being slated to adopt the system in 1873.

In general, such a message was well timed. The International Meter Commission reconvened in postwar Paris in the fall of 1872, its delegates having had time to think over the issues that had been brought up at the previous, war-aborted conference two years earlier. Broad agreement was reached with relative speed on the major topics, with the upshot being that a new standard meter and kilogram would be fabricated under the auspices of an international organization that would also house them.

The new American point man in Paris was Julius Erasmus Hilgard, who headed up the Office of Weights and Measures at the Coast Survey.

Born in the German Rhineland and a French speaker, Hilgard felt perfectly at ease with his European peers; he became a force in the proceedings and was elected to the twelve-member permanent committee charged with overseeing creation of the new meter and kilogram. The fabrication of standards and the execution of future comparisons, however, were going to take money and facilities, for which the commitment of foreign governments was necessary, and so in the fall of 1873 the committee requested that France call a diplomatic conference to negotiate a Treaty of the Meter.

As far as U.S. intentions went, Hilgard had noted to his foreign peers that there was strong American approval for an international bureau, and he believed its creation would be a "powerful means of propagation" of the metric system, noting that men of science had interest in the project. Eighteen such men—Hilgard and Joseph Henry among them—sent Frederick Barnard a letter asking the Columbia president to assemble an organization dedicated to international measurement reform. Barnard threw himself into the task and formed the American Metrological Society, which met for the first time at Columbia on the penultimate day of 1873.

At the meeting, Barnard was named president and John Kasson vice president, with Samuel Ruggles and Wolcott Gibbs among those on the Council, while Charles Sumner was in the first batch of those nominated and elected to membership. Although the primary object of the group was stated to be the improvement of "Weights, Measures and Moneys," the second was to help bring about universal standards in all areas of measurement, be it in temperature, energy, or the establishment of the prime meridian.

Still, the overarching goal was to put America on the road to metric adoption. However, a theory from across the ocean threatened to hamstring the American metric movement, as it already had the British one. This idea held that Anglo-Saxons didn't need the meter because they already possessed the perfect measure—the inch. And adherents to this theory had no less a personage than John Herschel to be their champion of it.

The Almighty Inch

For decades, John Herschel had been the preeminent scientific voice in the Anglo-Saxon world. His *Treatise on Astronomy* had helped to both popularize the profession in America and spark the observatory movement in the 1830s. By 1835 the Englishman's fame was such that his name was used to sell the greatest newspaper hoax in U.S. history, when the *New York Sun* claimed that Herschel had discovered a race of bat men living in huts on the moon.

Herschel also had a long involvement with weights and measures. He had been on the commission that decided how to restore the yard destroyed in the Parliament fire and had also held Isaac Newton's old job of Master of the Mint during the 1850s, during which time he put forward a plan for decimalizing the pound sterling. While not against universal measures per se—he was the first to suggest a universal time— Herschel was most definitely against the metric system, and he did more than anyone to turn British opinion against it in the 1860s.

His complaints against the metric measures were both philosophical and scientific. Though he derided the meter for being based on an inaccurate measure of a meridian, Herschel's greater point was that its being based on a meridian was a "sin" in the first place. Not that he was against using the earth as a standard of measure—he just favored another candidate.

The meridian's flaw was that there were an infinite number of them crossing the face of the irregularly shaped earth from pole to pole, and the choice of any one of them was egocentric. In contrast, there was a single axis that went pole-to-pole through the earth's core, and it was shared by all the people of the world. It was the polar axis, Herschel believed, that should be the universal standard of measure.

Prof. Davies had seized on the axis idea, and the Herschel paper he reprinted in his appendix outlined a plan to base a system of measure around it. In his own book, Barnard readily conceded that if the meter were done over again, it likely would have been based on the polar axis. To Barnard, however, it was a pointless argument—the meter was what it was and there would be no changing it now. Still, Herschel's idea convinced others, including the American W. W. Mann, who after twenty years of metric advocacy renounced the French system, publishing a book in 1872 that reworked his volume of the year before, swapping out the meter for the axis as the basis of the system he was advocating.

Most everyone who supported Herschel's idea had an ulterior motive of sorts, and that was how the inch could be mathematically derived from the polar axis. Twice the polar axis divided by a billion equaled 1.001 inches, an utterly negligible difference (about one quarter of the thickness of a sheet of paper) that made the old Anglo-Saxon measure, to the mind of Herschel and his followers, a more perfect "earth-commensurate" measure than the meter.

Though his own science was sound, Herschel had gotten the idea of the axis-based inch from a less-than-scientific source, a book called *The Great Pyramid: Who Built It and Why Was It Built?* by John Taylor. The publisher of Romantic poets like John Keats in his younger days, Taylor had become an armchair archaeologist with an evangelical bent, believing that the building of the Great Pyramid had been directed by God as a repository of His measures and made in proportion to His other great creation, the earth. Taylor contended that the Great Pyramid was not constructed using the cubit of the heathen pharaohs, but a longer one of 25 Anglo-Saxon inches, a standard he further believed had been used by Noah to build his ark and by Moses for the Tabernacle, making the inch a sacred measure.

While seeing little worth in the religio-archaeological aspect of Taylor's work, Herschel made the inch-axis connection the centerpiece of a paper he circulated to members of Parliament, which helped defeat the metric adoption bill. Of equal import, Herschel's imprimatur led one of his protégés, Charles Piazzi Smyth, to take a look at Taylor's theories.

A prodigy, Piazzi Smyth was appointed the astronomer royal of Scotland at twenty-six, and delivered his greatest contribution to the world while in his thirties with his pioneering of "mountain astronomy." In *Specialities of a Residence Above the Clouds*, Piazzi Smyth documented how he set up a temporary observatory at over two miles of altitude on the peak of Tenerife in the Canary Islands and proved its superiority over ones built at lower elevations.

In the mid-1860s, Piazzi Smyth became seized by Taylor's ideas, which the astronomer's bestselling *Our Inheritance in the Great Pyramid* put in a scientific context and spread to a much wider audience. It was at this point that Piazzi Smyth's life came to seem like a Victorian novel. To satisfy Taylor's dying wish, Piazzi Smyth went to Egypt to prove the old publisher's theories correct. Traveling to a Cairo that had been turned into a boomtown on account of the U.S. Civil War–induced Cotton Famine, Piazzi Smyth and his wife set up house in an abandoned tomb on the Giza plain. The Piazzi Smyths embraced the native customs, with Charles taking to sporting a tarboosh, a sort of soft fez that he would wear to the end of his days. Using state-of-the-art tools—some of his own invention—he and his wife performed the most comprehensive survey ever done of the Great Pyramid.

Piazzi Smyth published his findings in a three-volume work, which confirmed Taylor's and other discoveries and provided further ones, such as the year's length being spelled out in the 365.25 divine cubits that measured the perimeter of the Great Pyramid, and the sarcophagus of Cheops being a measure related to the English bushel. For Piazzi Smyth, the Great Pyramid was about more than archaeology and science—it was a physical complement to the word of the Bible.

If this all sounds bizarre, consider that little of what Taylor or Piazzi Smyth offered was new. That the Great Pyramid's size was related to the measure of the earth was a long-held view that had been confirmed by an archaeologist on Napoleon's fabled Egyptian expedition, who had also proposed that Cheops's sarcophagus was a standard of capacity, in keeping with the idea that temples were historically where measurement standards were kept. The recovery of ancient standards from monuments was the basis of historical metrology, and the length of the "sacred cubit" of the Hebrews measuring 25 inches came from Isaac Newton.

Piazzi Smyth's work struck a chord. It was received well by many of his fellow scientists, and its popularity spilled into Belgium and France. But his theories resonated with no one so much as the British Israelites, a growing movement that claimed the British to be a lost tribe of Israel. For them, the idea that the Old Testament cubit was based on the inch that the British had preserved down to the present day was a smoking gun for their beliefs. (Again, if this seems crazy, realize that the Book of Mormon chronicles an American branch of the ancient Israelites, a story Joseph Smith claimed to have found written in "reformed Egyptian" on golden plates buried in a hill in upstate New York in the 1820s.)

Piazzi Smyth's pyramid obsession lasted five years, to be replaced by a new one, spectroscopy, for which he returned to the life of a trailblazing scientist. He remained ever protective of his work on the Great Pyramid, however, and hostile to the French system that sought to depose its sacred measures. So when someone came along to assail his beliefs and research, he was only too willing to fight back.

Astronomical Flame War

In another situation, Charles Piazzi Smyth and Frederick Barnard might well have been friends. Both, after all, had chosen the same profession, and both were strongly religious. In a disagreement, however, the two were an incendiary match, with Piazzi Smyth being almost pathologically thin-skinned and Barnard borderline sadistic when it came to proving others wrong.

In *The Metric System*, Barnard called Piazzi Smyth's "attachment to the inch" near-fanatical and completely religious in origin, having nothing to do with science. Such charges greatly upset Piazzi Smyth, who believed that his investigations into the Great Pyramid in fact bridged the ever-widening gap between science and faith, by revealing God's true designs.

In a new edition of *Our Inheritance in the Great Pyramid*, Piazzi Smyth quoted Barnard's disparaging comments about him and refuted them, while calling the Columbia president the ultimate "pro-French metric champion," the French aspect of the system being for Piazzi Smyth, as with so many other Englishmen, among its most disagreeable qualities.

The aspect of the meter's Frenchness that most bothered Piazzi Smyth was its origins in the French Revolution, "the wildest, most bloodthirsty, and most atheistic revolution" there had ever been. He believed the metric system to be cut from the same cloth as Robespierre's Goddess of Reason, and he saw the decimal calendar with its obliteration of the Sabbath as proof of the link between measurement by tens and godlessness.

At the time, the Revolutionary era was far from ancient history, with a Napoleon having just been booted off the throne of France and the Terror having been reborn in 1871's Paris Commune, which saw the reestablishment of the Republican calendar and the senseless execution of the archbishop of Paris.

Charles Piazzi Smyth's views thoroughly nauseated Barnard. In a new edition of his own book, Barnard spent several pages quoting Piazzi Smyth's anti-Revolution rants, which he believed to be ample proof that Piazzi Smyth was out of his mind. Barnard held the meter up as the "innocent creation of human genius" and not deserving of being the target of Piazzi Smyth's vitriol. An Episcopal deacon, Barnard took specific issue with Piazzi Smyth's claim that the metric system had been birthed by atheists, pointing to a "Christian bishop" as having been its originator. (A laughable remark considering he was referring to Talleyrand, who didn't know how to perform mass the one time he attempted it.)

While Piazzi Smyth's rhetoric was over the top—as when he compared the meter to the mark of the beast—some of his criticisms were undeniable, such as his claim that the men pushing the meter were the same ones bent on "converting all the nations of the earth into one great people, speaking one language and using but one weight and one measure." If Hugo, Cobden, Sumner, and other internationalists had for decades been trumpeting the true importance of the meter and the franc as being steps toward a world without borders, how could they not be taken at their word? In an era that saw the united Germany and Italy consume nearly fifty states along with their marks of sovereignty such as weights, measures, and currency, it is understandable that the metric system represented to Piazzi Smyth the "anti-nation-existence metrological system."

The difference between Barnard and Piazzi Smyth came down, at root, to differing visions for the world. Barnard couldn't see how anyone in his right mind could see universal measures as anything but a good thing, while Piazzi Smyth saw them as the epitome of all that was wrong in a world that held Progress as its chief divinity. A diplomatic conference that sought to establish an official international meter could only make his stomach churn.

A Klean Kawse

"Never did so many Kaizers, Kzars, Kings, kum kling knit together in so Klean a Kawse to work so Kommendable a kure."

Thus did the humorist Josh Billings sum up the Treaty of the Meter, agreed to by seventeen nations on May 20, 1875. The treaty created a kind of metric embassy in the village of Sèvres just down the Seine from Paris, an estate that would physically be located within France, but extra-territorial. It would house the International Bureau of Weights and Measures, the organization now entrusted with the construction and administering of a new standard meter and kilogram.

The American Metrological Society was in the midst of its spring meeting when the treaty was announced, news made all the more sweet by the United States being among the signatories. The treaty electrified an organization that was already off to a running start in its little more than one year of existence.

Barnard had been circulating a petition endorsing the metric system and gotten an impressive array of signatures, including those of twenty-six congressmen and seventeen senators. A Boston member, J. Pickering Putnam, created another petition that asked his fellow architects to pledge to use the meter in their daily work starting July 4, 1876, and he was on his way to amassing hundreds of signatures.

But no one in the American Metrological Society was doing more than its youngest member, Melvil Dewey. The twenty-four-year-old Dewey was deeply impressed by Frederick Barnard and the organization he had created, with its excellent journal and the headlining names the Columbia president had recruited to the cause, which included Henry Wadsworth Longfellow, Oliver Wendell Holmes, Secretary of State Hamilton Fish, the now twice-failed vice presidential candidate Charles Francis Adams, the present and former New York governors Samuel J. Tilden and Horatio Seymour, and the landscape architects Frederick Law Olmsted and Calvert Vaux, whose Central Park had recently been completed.

Born in the year of the Crystal Palace, Melville Louis Kossuth Dewey grew up in the Burned-Over district of upstate New York, a place fired with zeal and passions like no other in the nation. It was here that Joseph Smith "translated" the Book of Mormon and the end-is-nigh Millerite movement began, and was also the home of such nonreligious activism as the first women's rights congress, held in Seneca Falls in 1848.

Formed of such missionary zeal, Dewey dedicated his life to reform at age sixteen, and he bought a pair of R-engraved cuff links to be a constant reminder of his choice. He would soon settle on three causes: the metric system, free libraries, and simplified spelling.

It was Dewey's interest in libraries that would make him famous, a success directly linked to his decimalist fervor for the metric system. While employed at his college library in Amherst, Dewey had a revelation. "I jumpt in my seat and came very near to shouting 'Eureka!' Use *decimals* to number a classification of all human knowledge in print."

The first edition of the Dewey Decimal System was published in 1876, a year in which everything came up aces for Dewey. In addition to his classification scheme, Dewey had a leading hand in creating the American Library Association and the *Library Journal*, as well as organizations and journals for the metric and spelling reform causes. He was secretary of all three societies and editor of all three periodicals.

Dewey founded the American Metric Bureau on July 4, 1876, with the support of Barnard, who served as its president. It was seen as a complement to the more high-minded American Metrological Society, with Dewey's Bureau meant to be a grassroots organization focused on distributing educational tools and promotional material. One of its vice presidents was Charles Francis Adams, a quote from whose father fronted the first edition of the *Metric Bulletin* and proclaimed the French system to be "in design the GREATEST invention of human ingenuity since that of printing."

In the wake of the Treaty of the Meter and Barnard's efforts, Dewey believed that the metric system stood the greatest chance out of any of his reforms of being adopted. Certainly, it had a leg up on simplified spelling, which suffered from the problem of looking illiterate. Josh Billings earned a living imitating bad spelling (one of the easiest ways to get a laugh in nineteenth-century America), and it was hard to tell the difference between his satire and such writing of Dewey's as "Speling Skolars agree that we hav the most unsyentifik, unskolarli, illojikal & wasteful speling ani languaj ever ataind." The *R*-cuff-linked reformer had nothing if not the courage of his convictions, though, and on his twenty-eighth birthday he legally changed his name to Melvil Dui.

By the end of 1877, the American Metric Bureau had sent out more than half a million pages of flyers and placards on the metric system. Dui was selling meter sticks, metric measuring tapes, sets of gram weights, charts, textbooks, and more, or just giving the stuff away to spread the word. He also got libraries to adopt standard metric sizes for their card catalogs.

Education remained a primary focus also at the American Metrological Society, which lobbied hundreds of colleges to make knowledge of the metric system a prerequisite for admission. But petitions and awareness raising went only so far; for the metric system to become an American reality, government intervention would be necessary.

Metric advocates were divided over how to make this happen. Some wrote memorials to Congress demanding that national conversion be mandated straightaway, but a more gradual approach was preferred by most, Barnard included. Their main priority was getting the government itself to go metric in areas like the post and customshouses.

To test the waters, a House resolution was introduced in 1877 that asked the departments of the executive branch to report what objections, "if any," they had to making the metric system obligatory for all government transactions, and whether and when it should be made the sole legal system for individual business. For however positively such questions were framed, all the reports save one came back negative, the most surprising of which was authored by J. E. Hilgard, the only American whose metric credentials surpassed Barnard's. In addition to being the head of weights and measures at the Coast Survey and on the American Metrological Society board, Hilgard had recently turned down an offer to become the head of the International Bureau of Weights and Measures at Sèvres.

When it came to the government, Hilgard's position was that the meter was being properly used at the Coast Survey and the mint, where it didn't come into contact with the public. To use it in an area like customs would be a mistake, because unfamiliarity with the system would lead to error. Only "by zealous inculcation, by agitation, by instruction in all public schools" could the meter become the choice of the people, and only then should it be made compulsory.

"It has ever been the practice of the Anglo Saxon people to make laws in conformity with customs," the foreign-born Hilgard wrote in his report, "not to create customs by compulsory laws."

To that point, the metric system had succeeded solely in countries headed by a king, kaiser, or emperor, and only by force of law. And not necessarily even then. The adoption of the metric system by Brazil in 1873 had gone disastrously, with the Quebra-Quilo ("kilo breaker") revolt, an event that began with the smashing of metric scales and snow-balled into a general insurrection against the central government that lasted months.

Metric advocates were far from discouraged, however—the legislative push was only just getting under way. In the meantime, however, Barnard and the American Metrological Society took hold of another reform that would be their legacy: Standard Time.

11/16

STANDARD TIME

A S PROMISED, the American Metrological Society had never been a one-issue organization. Pages of the AMS's journal show its members engaged in a wide variety of reforms, with the group's president pushing nothing so much as universal coinage. Unsurprisingly, Barnard opposed his friend Ruggles's plan, preferring a "metric dollar" of 1.5 grams of gold. Adopted in Japan and Argentina, the metric dollar gained some traction, but not enough. As ever, the real problem was that currency had been poorly integrated into the metric system in the first place. The same could be said of time.

Time was a hot topic in the 1870s. More than any other area of measure, it had been transformed by the industrial age. To see how, compare the footloose journey chronicled in Jules Verne's 1873 *Around the World in Eighty Days* to the ordeal endured by the first men to circumnavigate the globe.

Ferdinand Magellan had set out to accomplish what Columbus had failed to do—reach Asia by sailing west from Europe. Columbus had believed the earth to be far smaller than it was and so did Magellan, who wasn't prepared for anything like what awaited him after he passed through the straits later named for him.

After crossing an enormously wide ocean he didn't even know existed, Magellan finally reached Asia in the form of the future

Philippines, the inhabitants of which promptly killed him. Eighteen of the original 260 crew members would go on to complete the first circumnavigation of the globe in Magellan's name, which took almost exactly three years to complete. By these standards, to do it in eighty days was a miracle.

When Verne wrote his novel, such a lightning-fast round-the-world journey had only just become feasible, with the conclusion of three grand projects over a seven-month span: the Suez Canal, the Trans-Continental Railroad, and the Great Indian Peninsula Railway. Canals, railways, steamships, locomotives—all the wonders of the modern age now linked the world in a single interconnected circuit.

The plot of *Around the World in Eighty Days* revolved around Phileas Fogg's race against time to collect a £20,000 prize put up by his fellow members at the Reform Club of London. To win, Fogg would have to travel 4°30' of longitude per day, which, if measured as a portion of the equator, would equal 311 miles. The sun would rise an average of 18 minutes earlier every place he woke up. Too little to cause jet lag, but enough to notice.

For how dissimilar they appear, these measures—4°30', :18, and 311mi—are different ways of expressing the same thing. They all divide a sizing of the world by eighty: 360°, 24 hours, and 24,901.55 miles. The correlation of these measures had been a fundamental aim of the metric system, but its creators were inconsistent in their method; they split the earth into 400° and its circumference into 40 million parts, but divided the day by ten. A plan to correct this original error was proposed to the American Metrological Society by one of its most fiercely pro-metric members, Frederick Brooks, who advocated splitting the day into 40 parts.

Consider how easy it would be to calculate Phileas Fogg's average progress in a 400°, 40-hour, metric world, and how pleasing the result. His 30°-:18-311mi daily trek would be transformed into one that took

5°, 0.5 hours, and 500 km.★ It gets even better when you realize that all three measures would be called the same thing: 0.5 decagrade. In early versions of the metric system,† a *grade* equaled both one degree of circumference and 100 kilometers. In the AMS journal, Brooks proposed making ten of these—a decagrade—the new "hour." It was an incredibly perfect system, and, if instituted, anyone looking at a globe would instantly understand the interrelation of geography, time, and distance.

Of course, it would never happen. The 24-hour day was becoming a global standard faster than anything else, spreading as it did with the technology of the clock, a machine that came with the hours as its operating system. In fact, no other area of measurement was nearly so close to universal, a fact that remains true today.

Not that time didn't need to be fixed. It did, for a reason linked to the record-shattering speeds that steam power and titanic infrastructure projects had allowed.

Until the nineteenth century, people told time locally, calculating noon according to their own meridian—literally, to where they stood, making time the most self-ish of measures. In this world, Newark was one minute behind New York, Philadelphia five minutes behind, and Boston twelve minutes ahead.

It had also forever been true that you couldn't go anywhere fast enough for time differences to matter. At the average American latitude, it takes a day's walk of nearly fifteen miles due east to get to a place where the sun rises a minute earlier than where you started, an imperceptible

---

★   One-eightieth of the earth's circumference is approximately 500.94 km. The extra .94km owes mostly to the earth's oblate shape, making the equator slightly longer than any intersecting meridian.

†   Originally, the earth was a part of the metric system of length. It was divided into four quadrants, with the grade being the hundredth part of a quadrant, and the meter the hundred thousandth part of the grade. See table, p. 79.

difference considering sunrise can vary twice that much from one day to the next. Train lag, however, became unavoidable when a traveler found the sun rising nearly an hour earlier in New York than it had in Chicago the morning before.

The speed of the steamboat and train, by compressing distance, caused local times to collide. At any given stop along the Chicago–New York route, people would board the train carrying different times on their watches. The lack of standardized time was a problem apparent to any modern traveler.

The answer in most countries was simple. A single railroad time was set for the entire nation, often calculated at an observatory in the vicinity of the capital city. While traveling, a passenger was on national time; when he arrived at his destination, he switched to local time.

This worked splendidly in Britain, a long and skinny island, but in the continent-spanning United States, putting all railroads on Washington time would've meant true noon occurring at 3:02 P.M. on the watches of San Franciscans. Also problematic in the United States was the free market. Whereas state railway monopolies were the rule in other nations, America had hundreds of separate railroad companies.

The capitalist solution was simple—each company kept its own time. This was fine for commuters using a single line, but created existential agony for those making connections. Traveling salesmen had to rely on thick railway and steam navigation guides, just as Americans once needed almanacs with currency-exchange tables. The cover of the June 1870 *Traveler's Official Guide of the Railways and Steam Navigation Lines in the United States & Canada* displayed the problem neatly, with a central clock face declaring WASHINGTON noon, against which was marked the relative time in a hundred different travel destinations, from nearby Baltimore (12:02) and Richmond (11:58) to distant Vancouver Island (8:44) and Lima, Peru (also noon).

What made the situation truly maddening was the haphazard way towns chose which time to use. Some stuck with natural time, others chose the railroad time of nearby train stations, while still others picked the time of the nearest big city. In the 1870s, clocks up and down the same meridian (which by definition should have shared the same time) were put into all kinds of conflict with one another.

There had to be a better way. In 1879, the American Metrological Society delivered one, in the report of their committee on Standard Time, written by one of the nation's premier scientists—America's first weatherman, Cleveland Abbe.

Probabilities

That the better way involved a system of time zones will come as no surprise to us today, but realize how far from obvious such an answer was in 1879. It was one thing to swap local time for that of the railroad or the big city, quite another to set it to a geographical abstraction based on the location of an observatory in a foreign country (namely Greenwich) and then to say that hourwide zones of people should share the same time. Neither idea originated with Cleveland Abbe—at least two other people had independently come up with the time zone concept—but his "Report of Committee on Standard Time" set in motion the events that would lead to the system of universal time the world has today.

Like so many other important American scientists, Abbe started out at the Coast Survey, working there during the Civil War. Still in his twenties at war's end, Abbe apprenticed for two years at Pulkovo Observatory in St. Petersburg under the legendary astronomer Otto Struve, and then landed the job of running the Cincinnati Observatory atop Mount Adams.

At Cincinnati, Abbe focused on meteorology. Then one of the hottest fields in science, meteorology was a discipline that required tracking the most dynamic, chimerical phenomena of the natural world. Among the most enduring scientific undertakings of Joseph Henry's Smithsonian Institution had been the making of weather maps, put together from observations made by a countrywide force of amateur observers who telegraphed to Washington conditions like wind direction and speed, temperature, and barometric pressure. The accumulation of these maps revealed patterns that allowed Abbe to begin issuing the first American weather reports that sought to predict the weather rather than merely record it. In 1869, newspapers in several cities began running Abbe's "probabilities."

If weathermen frustrate us now, Abbe must have seemed like a sooth-sayer then. Storms, floods, and frost that had previously come without warning could suddenly be prepared for. The government saw the strategic importance of forecasting—for agriculture and shipping in particular—and started what would eventually become the Weather Bureau, which Abbe moved to Washington to help lead.

Tracking atmospheric events was nothing so much as a logistical issue, and discrepancies between his reporters using local or railroad time proved troublesome for Old Probs, as the not-yet-forty-year-old Abbe had come to be known by co-workers. For him, it wasn't just the railroad but the telegraph that was leading to difficulties.

The telegraph had revolutionized time in many ways, including its delivery. Time was no longer a thing locally sourced, to be puzzled over by a timekeeper with a sundial and unavailable on cloudy days. Instead, it could be purchased from Western Union for annual subscriptions costing as little as $12.50 for a weekly signal and as much as $500 for a daily one. A signal originating from the Naval Observatory arrived in the form of Washington noon, which would then be converted locally, for instance to 11:48:12.08 A.M. at the Pittsburgh post office. The telegraph could be used to automatically trip a public signal, such as cannon fire, the ringing of the city fire bells, and—most popularly—the dropping of a time ball, a relic of which ushers in the new year at Times Square.

Abbe's plan called for all the time balls in the county to be triggered by the same noon. Ideally, Abbe would have liked to use one time for the whole country, if not the world. Such an idea went back fifty years to when John Herschel first proposed "equinoctial time," and it had recently been promoted by Cleveland Abbe's Prusso-Russian mentor, Otto Struve. The most obvious candidate for a single global time was that of Greenwich, since it was already being used in a universal way.

Greenwich time was also known as ship's time, being kept by nearly all vessels at sea on account of the dominance of Britain's *Nautical Almanac*. The almanac contained the most widely used global positioning tool of the day, ephemerides, which were tables that gave the position of celestial bodies at any given time, essential to calculating longitude.

For however much ship's time was a familiar concept, the idea of instituting a single national time was too radical a step for the general public all at once. Abbe did want to introduce one, but only for use in train schedules and telegram time coding. To help familiarize people with "Railroad and Telegraph Time" (which he set to Greenwich +6 hours), Abbe suggested it be put on display in public buildings such as post offices and customshouses. Ultimately, however, his purpose was that it be the time used by all Americans for all things; as a way of making that happen, Abbe proposed the intermediary step of time zones.

Although they had clean longitudinal edges rather than jagged borders and Eastern and Central Time were called Atlantic and Valley Time, Abbe's zones were more or less the ones we know today. Everyone in America would relinquish time as they presently observed it, except for those lucky few who happened to live in cities such as Memphis, New Orleans, and Denver, which stood at one-hour (15°) intervals away from Greenwich and would be smack-dab in the middle of their zones.

Abbe and Frederick Barnard hoped for a speedy adoption of this two-tiered scheme, which was printed along with Abbe's report and distributed far and wide. It was a plan specifically aimed at solving the time problems of North America; Abbe and Barnard, however, soon came across a man with a time zone scheme he wanted to spread across the entire planet.

The Day of Two Noons

Although born in Scotland, Sandford Fleming had grown up to become one of the leading figures of British North America. While still in his early twenties, Fleming had designed Canada's first stamp, the 1851 "three-penny beaver," but his greater renown came from being the engineer in chief of his adopted country's answer to the transcontinental railroad, the Canadian Pacific Railway.

In 1878, Fleming published a plan to fix time globally. Fleming's concern was for the Phileas Fogg-type world traveler, constantly setting and resetting his watch when confronted with a new form of transport, be it steamer or train. While times within the states of Europe were less problematic than in the continent-spanning nations of North America, *inter*national time was an utter mess, as was one particular aspect of the clock itself, at least to Fleming's mind.

Fleming wrote that he had come upon the issue of time reform in 1876, when on a summer vacation in Ireland he found himself stuck on a railway platform waiting for a 5:35 P.M. train to arrive that never did. Not because his watch was wrong—the great fear of travelers—but because of a misprinted schedule reversing P.M. and A.M. The experience (or rather, the idea of such an experience—it seems the anecdote was a fib) led Fleming to become a lifelong enemy of the A.M./P.M. division of the clock.

Originally, two sets of hours existed because days and nights were divided separately. The *meridiem* or noon occurred at the middle hour of the day, which is why the midday break in Spanish is called the *siesta*, for the "sixth" hour; similarly, to do something at the eleventh hour was to do it right before sunset, in the final twelfth of the day. The mechanical clock ushered in the practice of restarting the hours at noon, hence the creation of A.M. and P.M. (for *ante* and *post meridiem*, before and after midday). Fleming considered the double-counting of hours to be utterly illogical, like negative numbers on a thermometer, and preferred numbering them 1 through 24—but for local time only. He also advocated that clocks keep a second, universal time that eliminated numbered hours altogether.

To achieve this, Fleming planned to set up time zones across the planet. Like Abbe, he wanted to use one-hour offsets of a meridian such as Greenwich, but foresaw the problem of no choice being neutral. To Fleming's mind, a true universal time would cause people to think of themselves as cosmopolitan, a "world citizen" literally. To effect this, Fleming assigned each time zone a letter, with the letters becoming the new hours. Once it struck noon in a given time zone, it was that time zone's hour throughout the world. So when the sun crossed the meridian over Memphis in the middle of the S zone, it would be S:00 everywhere; an hour later, when noon hit Denver in the T zone, it would be T:00 in Paris and Tokyo too. For convenience, Fleming created a dual-dial timepiece so people could adjust a set of numbered hours to overlay the lettered ones, but his hope was that local time would fade away and the world would just be left with "Cosmopolitan Time."

Although lettered hours would prove to be too radical, Abbe and Barnard were taken with Fleming's plan, particularly the idea that time zones were not a way station to a universal time but the permanent solution. They made common cause with the energetic Fleming, who joined the American Metrological Society in 1880 and would eventually give up on lettered hours, if not the idea of a 24-hour clock. The three of them spread out across the Western world trying to sell Standard Time.

The Washington-based Abbe made fast progress on the home front, with the nation's astronomers and other scientists lining up behind the time zone idea. Barnard and Fleming traveled to Europe to pitch the international congress scene, but here their scheme fell on deaf ears.

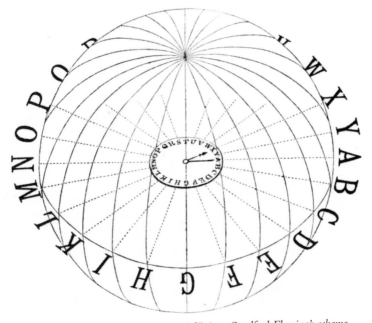

*Called "Cosmopolitan" or "Terrestrial" time, Sandford Fleming's scheme
proposed replacing the numbered hours with lettered meridians. Exactly which
letters would be applied to which meridians shifted and were debated, but
members of the American Metrological Society embraced Fleming's idea.*

One roadblock to selling Standard Time in Europe was that it solved
a problem that in most places didn't exist, Phileas Fogg–like travelers
aside. National times worked too well. Why on earth would France turn
their clocks back nine minutes to get on the English hour? Especially
considering that Greenwich time wasn't even used in British-ruled
Ireland, twenty-five minutes behind on Dublin time. As Fleming had
foreseen, not being neutral was part of the problem. Some saw agreeing
to Greenwich time as tantamount to accepting Greenwich as the prime
meridian of the world, a tough pill to swallow for some in the non-
Anglo-Saxon world.

Back home, however, the trio had struck gold with William F. Allen.
Another former railway engineer, the youthful Allen was the editor of

the *Traveler's Official Guide* and—more crucially—head of the Railway Time Conventions, which coordinated the country's railroad schedules.

Although the railroads had considered and rejected a time zone plan a decade earlier, Allen believed that such reform could no longer be held off. Allen put the fear of god into the railroad companies, convincing them that if they didn't do something fast the government would, and a Standard Time based on smoothly defined zones of longitude such as Abbe had drawn would create a scheduling nightmare. To head off such an eventuality, Allen created a scheme that looked to salvage the system of one time per railroad line as much as possible, leading to his time zones looking like pieces of a jigsaw puzzle. They included an enormous central time zone that touched the Atlantic Ocean at Savannah and reached all the way west of Denver to include El Paso, places that were naturally two hours apart. Still, it adhered to the general principles of Abbe's plan and constituted a huge improvement over the prevailing state of affairs.

Barnard, who was becoming increasingly frustrated by the glacial pace of representative government and its lack of proactivity when it came to the metric system, discovered big business to be a different beast entirely. Allen's plan, proposed at the April 1883 Railway Time Convention, won approval at its October meeting, with nationwide conversion set to take place the following month on November 18, 1883, a Sunday which came to be known as the Day of Two Noons.

Habituated as we are to changing clocks for daylight saving time or after a flight, we think of time as something that can be manipulated. But to Americans of 1883, most of whom had grown up understanding time to be a local phenomenon of the sun, the idea of noon repeating itself on a brisk autumn weekend was supernatural. Equally radical was the idea of church bells around the country ringing the hour in unison, rather than unrolling like a carpet across the land.

Standard Time was embraced not just by railroad companies but most American cities on the Day of Two Noons, when eighty of the nation's hundred largest cities also changed their local times. The reaction from the public was by and large positive. Most tangibly, the pocket watches of rail travelers were now set to the same minute no matter what station they had boarded from, and timetables all at once became comprehensible.

With America won, the creators of Standard Time looked to spread their time zones abroad. But before that could be done, the thorny issue of the prime meridian would have to be settled. Two weeks after the Day of Two Noons, the U.S. secretary of state sent out invitations for an event to be held in Washington the following fall: the International Meridian Conference.

The Starts and Ends of the Earth

That a prime meridian could be a matter of opinion has to do with the fact that longitude is relative; lines of latitude (or parallels), on the other hand, are absolute. So Jefferson could agree with Talleyrand that a pendulum should be calibrated at 45°N latitude, Canadians understood perfectly the meaning of the Young America slogan "Fifty-Four Forty or Fight!" and 0° latitude—the prime parallel of the world—is known to all as the equator.

A prime meridian marks 0° longitude, but whereas the equator is the one and only midway point between the poles, any meridian can be chosen to stake where the world starts and ends. For many kings, the appropriate prime was the meridian on which their royal observatories were located. At the beginning of the century, Laplace had called the idea of national prime meridians absurd; he would have been sad to see how little improved the situation was by 1884, with 14 different prime meridians remaining in use. On a French map, Paris may have been 0° longitude, but it was 2°E on a British map and 79°E on American ones. Geographers and geodesists the world over found the situation intolerable and a major impediment to cartography, while men like Victor Hugo mentioned a universal prime in the same breath as a universal language and currency.

It also had bedeviled attempts to sell Standard Time in Europe, having come up at the International Geographical Congress that Sandford Fleming attended in Venice and the international law congress that Frederick Barnard went to in Cologne.

If a meeting of lawyers sounds like a funny place to be pitching universal measurement reform, realize that the Association for the Reform and Codification of the Law of Nations was in many ways the successor to the old peace congresses, having two of same men behind it, the Learned Blacksmith Elihu Burritt and the Belgian Auguste Visschers. The most crucial figure in the association, however, was a friend and collaborator of Frederick Barnard's, David Dudley Field.

Field was the Young America Democrat who had brought codification to the United States in the late 1840s with his Field Code of Civil Procedure. One of the most important legal reformers in American history, Field wanted to simplify and make more uniform the American legal system, trying to reduce its reliance on case history. His code or some part of it had been adopted by more than half the states in the union, and had been the critical influence in an overhaul of British jurisprudence that began in 1873 and was spreading to all corners of the Empire. If the Anglo-Saxon world had an answer to the Napoleonic Code, the Field Code was it.

The year 1873 had also seen the first Law of Nations congress and the publication of Field's *Draft Outlines of an International Code*, which attempted to create a universal legal system, a kind of holy grail to liberal internationalists. Several had taken stabs at it, including Leone Levi, the veteran universal measurement activist who also reemerged at the Law of Nations congresses. Crucially, a world legal code would form the basis for a court of international arbitration, which had remained the foremost desire of the peace movement since Cobden.

The sole outside contributor to Field's *International Code* was Frederick Barnard, who wrote the section on what the legally recognized measures of the world should be. Besides his usual arguments for the metric system and the 1.5-gram gold dollar, Barnard presented a reconfigured calendar in which months alternated between 30 and 31 days. He also tackled the issue of the prime meridian, arguing for a Greenwich standard chiefly on the basis of the dominant position of Britain's Nautical Almanac. But Barnard's opinion was just one of many.

By the time the International Meridian Conference opened in Washington in October 1884, the issue of the prime had been well ventilated over the previous decade or so. Many neutral meridians had been proposed, including Ferro in the Canary Islands, the choice of Ptolemy and still in wide use; the Peak of the Tenerife, which Charles Piazzi Smyth had scaled; Jerusalem, suggested for its centrality not just to three major religions but three continents; and the Bering Strait, because a line drawn through it crossed virtually nothing but ocean. One practical consideration, however, voided all of these candidates. A prime meridian had to be the astronomical reference point of the world, and none of these neutral sites had or could reasonably sustain a major observatory. In the end, what choice could there be but Greenwich?

The international consensus for the Greenwich prime had been building, and not only in the English-speaking world. The Prusso-Russian Otto Struve had been banging the drum for a Greenwich prime since at least 1870, and the cause had also been taken up by Adolph Hirsch, the Swiss astronomer who had become head of the International Bureau of Weights and Measures after J. E. Hilgard turned down the job. Hirsch loathed national times, and compared the benefits of a universal prime meridian to those of the metric system; he also pointed out that adopting the meter hadn't harmed countries who had given up their native measures, so why should adopting the Greenwich prime be any different?

The United States had been scheduled to have Frederick Barnard as its lead delegate at the Washington conference, but the educator pulled out at the last minute on account of conflicting matters at Columbia. This was to the relief of some, who believed a deaf man couldn't properly represent American interests, and the dismay of others, who recognized Barnard as an international authority on the matters at hand. Cleveland Abbe, however, would show himself to be more than adept at the verbal sparring that would prove necessary.

"America has never lost a war but never won a conference," Will Rogers quipped, but America won this one, so far as there was winning to be had. More surely there was a loser, and that was France.

For once, the French were on the wrong side of a universal standards debate. This mostly reflected the fact that in other cases theirs had been the standard internationalists had rallied around, while now the choice was a British one. The shoe was finally on the other foot.

France refused to even consider using a Greenwich meridian on their maps, let alone to tell time. The Paris meridian had a glorious past and France's naval almanac was second only to Britain's, if an extremely distant second. The French had much to lose—the Paris Observatory would need to redo its star charts, and its ephemerides business would vanish—and very little to gain except being a good world citizen.

The French had as their delegate Jules Janssen, the astronomer who discovered helium. Janssen had one essential point to make—that it did not matter what the prime meridian was, so long as it was neutral. Meaning: not Greenwich. As Abbe questioned Janssen over the meaning of the term "neutral," the exchange turned sharp.

Abbe asked rhetorically if the metric system were neutral, and answered that it was not—it was French, and its Frenchness would never change, a fact that did nothing to lessen its having been a boon to humanity. That being the case, Abbe challenged Janssen over the very concept of a "neutral system of longitude," claiming such a thing to be "a myth, a fancy, a piece of poetry," unless he could show precisely how it could be achieved.

Janssen took clear offense at Abbe, and before even addressing the issue of the meridian, felt the need to defend the neutrality of the meter. The metric analogy was hard to defend against, though, as it hoisted the

French on their own petard. Metric advocates had been chiding Anglo-Saxons for decades with the argument that the meter was nothing more than a useful tool, to be adopted like any other useful tool. The Greenwich meridian was already the de facto prime meridian of the world because its utility had been proven, and the only reason to reject it was national prejudice.

A neutral meridian a loser, France tried to horse-trade its approval of Greenwich for Anglo-American adoption of the metric system. This failing also, Janssen used the conference as a platform for decimal time and angles, putting forward a scheme that essentially matched the 40-decagrade plan published by the American Metrological Society. Although the idea went beyond the scope of the gathering, delegates passed a resolution calling for further study into the subject.

The conference ended with a resounding vote approving Greenwich as the world's prime meridian. In doing so, delegates also approved an antiprime—the "zero of Time"—or what would come to be called the International Date Line, solving another vexing issue that had first been realized during the Magellan expedition.

When they made landfall on the Cape Verde Islands off the Atlantic coast of Africa, Magellan's remaining crew had their first contact back with European civilization. For all the great relief that their ordeal was over, the men made a disconcerting discovery. They were told on shore that it was a Thursday, whereas on ship they had been quite convinced it was a Wednesday. Despite their scrupulous observance of the Sabbath and the other days of the week, they had somehow lost a day, which would come to be understood as a necessary consequence of a westward trip around the world.

The corollary conclusion—that one could gain a day by traveling east—provided the kicker at the end of *Around the World in Eighty Days*. It actually took Phileas Fogg eighty-*one* days to circumnavigate the planet, and he arrived in London believing he had failed in his quest, only to discover it to be a day earlier than he thought, and himself £20,000 richer.

Fantastic Freaks

The Meridian Conference solved an argument, but there was no immediate groundswell of nations adopting either a Greenwich prime or Standard Time. To be sure, not even in America did everyone love Standard Time. Cities like Detroit that were caught on the edge of a zone had their sunrises and sunsets seriously upset and would flip-flop over which time to keep for years to come. Other cities just got screwed, like Savannah, which was stuck in the central zone and no longer had a sunset much later than half past six, even in summer.

Then there were those philosophically opposed to the new, more artificial time. Claiming it went against divine instructions, one individual deemed the meridian congress as yet another example of "the dread international conference which transcends all mere radical politicians in seeking ever by blood and fire to destroy most completely the ancient and necessary barriers between the nations, and to form all mankind into one vast, headless society."

This was a typical statement from Charles Piazzi Smyth (who, for the record, favored the Great Pyramid as the prime meridian). But while the Astronomer Royal of Scotland had had enough of profane plans to reorder the world, there were others who had had quite enough of Charles Piazzi Smyth.

"It is a fact," Frederick Barnard wrote in the preface to 1884's *The Imaginary Metrological System of the Great Pyramid of Gizeh*, "that this strange fantasy has been adopted as the creed of a numerous and actively militant sect, who, not content with cherishing their favorite hallucination among themselves, have found in it the inciting motive of a crusade against the spirit of progress of the age."

Barnard relished the opportunity to demolish once and for all the pyramid theories of Piazzi Smyth and his followers. In a postscript to his book, however, Barnard noted that he'd need hardly have bothered, as a former disciple of Piazzi Smyth's, Flinders Petrie, had done the job better than Barnard ever could.

Piazzi Smyth had been a longtime friend of Petrie's father, and when Flinders was a boy, the astronomer had taken a particular shine to him, perhaps because the younger Petrie like himself was a prodigy. By the age of sixteen, Flinders was already an Egypt expert and assisting Piazzi Smyth in measuring casing stones of the Great Pyramid in a London museum. Soon Petrie was publishing his own books, which included support for Piazzi Smyth's work, and in 1880 he set off to survey the Great Pyramid in part to prove the correctness of the older man's theories, as Piazzi Smyth had once done for John Taylor. Petrie would spend three winters on the Giza plateau, embarking on a career that would revolutionize archaeology.

While in Egypt, he discovered a fatal flaw in the work of his mentor. Piazzi Smyth had used a baseline in his survey that was too low, which meant he had been measuring the foundation of the Pyramid, a revelation that made all of the astronomer's numbers fall apart. An American follower of Piazzi Smyth's who spent time with Petrie in Giza said, "Well sir! I feel as if I had been to a funeral."

For whatever pleasure the ever-combative Barnard took in seeing Piazzi Smyth's theories refuted, it hardly mattered. While Standard Time soared, hopes for American adoption of the metric system had sunk, with bill after bill dying in the Congress through the early 1880s. Credit for this was taken by Charles Latimer, one of Piazzi Smyth's most devoted acolytes and a man who cared not a whit what Barnard, Petrie, or anyone else said about the pyramid inch.

Another railroad engineer, Latimer had struck it semirich from the Witch-Hazel Coal Mine in the Alleghenies, which he discovered using a divining rod, an experience that only reinforced his mystical beliefs, which would find a focus in the writings of Charles Piazzi Smyth. In 1879, Latimer formed the International Institute for Preserving and Perfecting Weights and Measures, with the express purpose of keeping the meter—which he called the devil's work—off American soil.

The organization counted three members in its council: Latimer, Piazzi Smyth, and Joseph Wild, a minister who had found success preaching British Israelism in Brooklyn. The Israelite movement had grown stronger, especially among Americans feeling increasing kinship with Britain and alienation from the rest of Europe. While the exploits of the British Empire spoke to the superiority of the Anglo-Saxon race, the Continent was the source of revolutions, despots, anarchists, Communists, and teeming immigrants, now streaming across the ocean in the millions. Tying the metric system with the latter and customary measures with good old Anglo-Saxon values proved a winning combination. It was only followers of Darwin and the infidel who wanted to adopt "the fantastic freaks of the French Revolution."

Latimer came off like a deranged version of Piazzi Smyth, particularly when he complained about things like the Statue of Liberty having been constructed using millimeters, but when it came to lobbying, his organization proved to be at least the match of those of Frederick Barnard and Melvil Dui. The International Institute swamped Congress and the president with petitions and open letters, protesting the metric system, Standard Time, and the fact that American dollars were being used to support the International Bureau of Weights and Measures.

Barnard became convinced that Latimer was scaring off politicians, another group of people the Columbia president held in low regard, as did most Americans of the day. Government corruption was rife through the Gilded Age, even touching the cause of universal currency via the *Stella*. A four-dollar coin worth 20F, the Stella was part of a renewed effort by John Kasson to have the United States join the Latin Monetary Union. The scheme ended in a farcical scandal when the gold coins—specimens of which had been given to congressmen in the hundreds—began turning up as jewelry in brothels all over Washington.

The metric system had no such enticing baubles to offer, and Congress had done nothing to help the meter despite years of efforts on the part of Barnard and the American Metrological Society. Never one to take injustices lying down, Barnard railed against representatives who threw his petitions in the dustbin and responded to nothing but patronage. Politicians didn't lead but followed, he said, and were unwilling to do anything that the public wasn't already demanding. For Barnard, the ancient quandary of lawgivers was easily answered: Representatives should do what was wisest and best for the people, no matter what the people themselves thought they wanted.

Blaming venal politicians and conspiracy theorists like Latimer was too easy, however. After all, what sane politician would advocate metric adoption when the head of the Office of Weights and Measures was steadfastly against it? Those wanting the meter were a tiny elite, many of whom spent their lives in the ivory tower.

But it was in the ivory tower that Barnard thrived. In his midseventies, he had grown ever more focused on Columbia, which he was building into a world-class Academy of his own. To create a library worthy of such an institution, Barnard turned to his old metric compatriot, Melvil Dui, whom he also asked to head the world's first library sciences graduate school, which he wanted open to women, coeducation being another of Barnard's causes (making the future naming of a separate-education women's college after him decidedly ironic). Barnard's offer came with a catch, however, as he demanded that Dui change his name back to Dewey. His old mentor didn't want to see such an "eccentricity" held against him.

While his Columbia legacy was secure and growing, membership in Barnard's metrological society was dwindling, having shrunk to a third of its high-water mark by 1886. Barnard himself had already accepted that metric adoption in America wouldn't happen for another genera-tion or two, and at the annual American Metrological Society post-Christmas meeting, Barnard wondered aloud if it wouldn't be best to just shut the organization down.

"I am thoroughly opposed to allowing this Society to die," Melvil Dewey said, and rose to point out its accomplishments with Standard

Time and public education on the meter. He urged perseverance. Persevere it would, for a time, but Frederick Barnard himself would soon expire, dying at his home on April 27, 1889.

For one who fought so hard for the meter, Barnard died in a fitting year. After nearly twenty years of scientific squabbling and difficulties, a new standard meter and kilogram had finally been unveiled, relegating the standards Barnard had inspected during the 1867 Paris Expo to the museum shelf. It was one hundred years since France had cast off its old measures, and the French celebrated the centennial of their revolution with another exposition, this one fronted by the grandest entryway ever built—the Eiffel Tower.

The year 1889 proved that the fantastic freaks of the French Revolution were as influential as ever, with grand internationalist projects aplenty. That summer, Paris hosted the inaugural meetings of both the socialist Second International and the First Universal Peace Congress. The latter was chaired by Frédéric Passy, an ardent admirer of Cobden's who that same year cofounded the Inter-Parliamentary Union, which aimed to establish an international court of arbitration. Such a court was also among the goals of delegates meeting in Washington at the first International Conference of American States, the future Pan-American Union, which looked to start an American customs union and a north–south Pan-American Railway.

Not a single one of these utopian projects would turn out as its planners hoped, but all would be influential. Influential, too, would be the tools for progress that had been developed during the revolutionary era. While these were at first spread beyond France on the points of bayonets, it was their original promise that now resonated, particularly among new nations looking to join the modern world as they emerged from feudal empires, both on the edges of Western Europe and on the other side of the world.

$^{12}\!/_{16}$

or Three Quarters

A TOOLKIT FOR THE WORLD

T HE LIFEBLOOD of the meter and franc had always been new and reborn nations. Belgium, the first country to successfully go metric, did so as part of its break from Dutch rule, while France readopted the meter in the wake of a second revolution. Unification spurred metric adoption in Germany and Italy, while throwing off the Spanish colonial yoke was the prod for Latin American nations. The breakup of the Ottoman Empire created new nations in Greece and the Balkans, all of which saw the meter and franc as part of the transition from a medieval to a modern society. Such a leap was easier said than done, and the only country that proved capable of making the move directly wasn't from Europe, but Asia.

In one of the most remarkable turnarounds in history, Japan had gone from a land of medieval feudalism to a modern industrial power in two generations. Closed for centuries, Japan was "opened" in 1853 by Admiral Perry and his black ships in what was America's first foray into throwing its weight around across an ocean. Far from being forced from abroad, however, the opening to the West was driven from within, in particular by Japanese who noticed how badly China had fared in trying to shut out the Europeans, with little more than repeated humiliations in the Opium Wars to show for it.

To be sure, the transition wasn't easy or bloodless, as the shoguns—a far more brutal lot than the French seigneurs—did not give up power readily.

But once the nation united behind restored imperial rule, it was remarkable how methodical its leaders were in looking abroad for how to transform even the basic ways of society. There was a certain tradition of this, considering that a foreign country, China, had been the source of Japan's numbers, weights, measures, coinage, hours, calendar, zodiac, written characters, religion, philosophy, and even its name for itself. ("Japan" or "Nippon" comes from a word referring to the sun's rising, a geographical reference as made from the mainland.) In a way, the Japanese just began looking to a different provider for the same tools. And they didn't merely copy standards and technology, they considered everything the West had to offer, including religions and languages. Even the Western diet was largely adopted, and a land of beef-averse Buddhists turned rapidly carnivorous.

The Japanese began using Western numerals—the ones we call Hindu-Arabic—as a prerequisite to assuming its other measures, which the island nation swiftly did. In the early 1870s alone, Japan adopted the 24-hour clock, the Gregorian calendar, and decimalized Western-style coinage. They became stalwart attendees of international congresses and took to heart their resolutions more than anyone else, becoming, for instance, the only nation to adopt the Greenwich prime and Standard Time directly after the meridian conference, even though the island nation had little issue with longitude and was far removed from Britain and North America.

Japan's most effective modernizing policy was its employment of *oyatoi gaikokujin*, foreign experts brought to the Land of the Rising Sun for brief teaching stints. They were to impart what they knew, be paid handsomely, and leave. This policy put Japan in marked contrast to China, where the emperors relied on foreigners to keep their vast collections of clocks and other automata going, or Russia, where the language of science remained German.

As the West was not a monolith like China, it didn't generally present single solutions. Japan comparison-shopped sources for its 1889 constitution and legal system, passing over liberal American and French models in favor of the appealingly autocratic Prussian one. Germany's own remarkable ascent proved most instructive for Japan, who followed the German lead in reorganizing its army, and also in what to do after you win a war.

Japan had come into conflict with China over Korea, and they shocked the world by crushing the neighbor that had so long dwarfed it, transforming their island nation from a good little student of the West to a force to be reckoned with. In victory, the Japanese pulled the same trick on the Chinese that the Germans had on the French, extracting large reparations payments that allowed them to go on the gold standard, becoming the first Asian nation to do so. Also as was the case with Germany, this was done without adopting the franc, a move in which they were going against the grain.

By the 1890s, coins set to the franc were being used in more and more nations around the globe, despite the fact the Latin Monetary Union kept rejecting candidates, nixing Serbia alone three times. Countries, however, could unilaterally join the franc zone simply by aping Latin Union standards. Much of South America had done this, as had a trio of former Ottoman nations who chucked the Sublime Porte's lira for the franc-derived leu (Romania), dinar (Serbia), and lev (Bulgaria). Ditching the archaic standards of an overlord was also the thinking of the duchy of Finland when it exchanged the ruble for the markka, which it set to equal one gold franc. Soon enough, however, Russia created a gold ruble that it too pegged to the franc.

For Finland and the ex-Ottoman countries, the choice of the meter was just as obvious as that of the franc, as it was the main standard of "civilized" Europe, a club they desperately wanted to be a part of. From Japan's vantage point in the Pacific, however, the franc seemed underwhelming, especially compared to the dollar, which had remained the universal coin of trade in Asia. The meter seemed similarly remote, as the main Western powers in the Pacific—the United States, Britain, and Russia—all used the same foot. Most of East Asia had an equivalent measure, with Japan's *shaku* being particularly close in size to the English foot. On the other hand, the Japanese were well aware that there were those in the Anglo-Saxon world who wanted to adopt the metric system. And in the 1890s, it had begun to look as though they would get their wish.

A Metric Fallacy

On the second day of 1890, President Benjamin Harrison broke the seals on a shipment that had just arrived at the White House from France. It contained meters number 21 and 27 and kilograms number 4 and 20, prototypes that had been randomly selected by the International Bureau of Weights and Measures for the United States as per the Treaty of the Meter. At least one witness to the event, Thomas C. Mendenhall, badly wanted to put these new standards to use in America. Having just become Superintendent of the Coast Survey—and thus the man in charge of the nation's measures—Mendenhall was uniquely in a position to do something about it.

In earlier days, Mendenhall had been one of the *oyatoi gaikokujin*, going to Japan in the late 1870s to chair the physics department at Tokyo Imperial University. Shortly after returning to the States, he went to work for Cleveland Abbe at the weather service and in 1885 joined the American Metrological Society, where Mendenhall found men who shared his intense dislike of traditional measures and all their eccentricities.

From his start at the Coast Survey, Mendenhall worked behind the scenes for metric adoption. First, he got the secretary of state to propose a resolution at the Conference of American States that all its members go metric, a move most obviously targeted at his own country. Next, he convinced the secretary of the treasury to present bills to Congress for making the metric system obligatory at the customshouses. In 1893, Mendenhall made his boldest move, releasing an order that switched the "fundamental" standards of the United States from the yard and pound to the meter and kilogram.

In a stroke of the pen, the foot went from being defined as a third of the imperial yard in the United States to 1200/3937; (0.304801) meters. The change was on the one hand radical, and on the other, common sense. The metric prototypes were the most accurate in the world and America was paying good money for them besides, so why not use them? The Japanese had essentially made the same decision, setting their shaku to the minimally smaller 1200/3960 (0.3$\overline{03}$) meters.

The Mendenhall Order was one of many significant acts taken by the Coast Survey in defining the nation's measures; it would also be its last. In 1901, responsibilities for U.S. weights and measures were shifted to a full-fledged Bureau of Standards, which replaced what had been a small office with a full-scale agency and physical science research laboratory, the first run by the federal government. The new organization was placed under the directorship of Samuel W. Stratton, who like Mendenhall was a physics professor from the Midwest. At the time, it looked like Stratton's duties would include presiding over America's transition to the metric system.

Momentum had been building in Congress since Mendenhall's initial push. In 1896, a bill to take the country metric actually passed the House. The success lasted mere minutes, however, as the bill was pulled back by a motion to reconsider once some of the nays convinced some of the yeas that angry farmers bewildered by having to chuck their bushels for liters would show their wrath in the upcoming election. The measure got tabled. Unvoted-upon bills and congressional reports began to pile up, but in a way that suggested progress, so much so that by the end of 1901 sponsors believed the time was ripe for another vote—in fact, they felt they had a sure winner on their hands. Congressional hearings were set up over two bills, both designed to take the country metric all but immediately, and the system was put on its most public display yet in America.

The sessions were conducted like a victory lap. Dozens of witnesses sang the meter's praises, including George Westinghouse, who had recently bested Thomas Edison in the AC/DC "War of Currents," and

Lord Kelvin, one of the world's most respected scientists, who had been fighting to have the metric system adopted in his native Britain. It wasn't the first time Kelvin had spoken in the United States on the subject, having previously apologized to the American people for his forebears' having saddled them with such dreadful weights and measures. "I look upon our English system as a wickedly brain-destroying piece of bondage under which we suffer."

With the public airing and imminent congressional vote, opposition to the meter gathered swiftly. A scathing *New York Times* editorial complained that the metric system was being rammed down the country's throat with scant regard for a cost that would exceed the market cap of U.S. Steel. The meter was legal for anyone to use and "it is because so few want to that the advocates of the system are impatient." As for who was pushing so hard, the *Times* said it was those who would benefit most from it and wagged an accusatory finger at Stratton and the "specialists" at the new Bureau of Standards.

The organization became a lightning rod of criticism and Stratton a villain to metric opponents, who didn't like the fact that the bureau was both a government agency and a metric advocacy group. The fit was indeed awkward. Stratton served as technical adviser to the House committee and not only helped put together its list of metric pom-pom-waving witnesses, but also questioned them during hearings. For Stratton's part, he simply wanted America to have the best system of standards in the world, and he believed the country was being held back only by "a little coterie of ultra-conservatives."

The coterie Stratton had in mind was a pair of editors of industrial trade magazines—Frederick Halsey of *American Machinist* and Samuel Dale of *Textile World*—whose antipathy toward the metric system and Bureau of Standards chief would bind them together for much of the next quarter century.

Halsey and Dale believed that the meter was being pushed against the will of the people by a scientific elite using underhanded tricks to get their way. Their prime example was the Mendenhall Order, which they contended violated the section of the Constitution that gave Congress the power to "fix the Standard of Weights and Measures." Halsey and Dale attacked the idea that the meter was inevitable, maintaining that it wasn't widely used in the countries that had ostensibly adopted it. On the one hand, their point was spot-on; places like the Ottoman Empire, Latin America, and Japan were claimed as being in the metric fold when such was not the case. But Halsey and Dale further stated that even in Western Europe the metric system was not universal, and that its introduction had wreaked havoc in industries. This was true, but only in instances. When it came to textiles, the meter still had yet to fully replace the old measures, creating a Continental Babel that profited Anglo-Saxon producers, whose lead in manufacturing had set the industry standard. But although even France had regions that continued to use customary measures, the idea that Europe was anything other than well on its way to full metrication was faulty. The more valid point was that conversion was not as easy as was being claimed.

Congressional sponsors decided to wait for the Halsey-Dale storm to pass and pulled the metric bill for clearer days, but the two editors only got better organized, lining up engineers and industrialists who backed their opinions. In 1904, they published *The Metric Fallacy*, a compendium of their objections that as much as anything ever printed kept the United States from going metric.

The argument was not closed—Alexander Graham Bell provided the best pro-metric congressional testimony yet in 1906—but the fight was lost. Beyond Halsey and Dale, the metric push of the nineteen-aughts failed because the main argument of advocates—that the meter was inevitable and the United States faced dire consequences if it didn't get on board—just didn't jibe with how Americans were seeing themselves at the start of the twentieth century. The evidence didn't bear it out either. The United States was exporting not only its products in record amounts but its ideas and power. Indeed, the country was evolving into an empire, with the annexation of Hawaii and the takeover of the Philippines and Puerto Rico after the Spanish-American War. America was emerging as one of the strongest nations in the world, fulfilling the massive potential so long seen for it.

Why should such a nation take a backseat to anyone? More pleasing arguments were being made by reformers who wanted to *spread* Anglo-Saxon standards, not discard them. And on the top of their list was the English language.

Speling Skolars Agree

For many Americans, the Spanish-American War was a welcome entry into the imperialist game, while for others it stood as an appalling betrayal of all the country stood for. The American Anti-Imperialist League was formed by those in the latter camp, and boasted a membership that included such luminaries as Mark Twain, Samuel Gompers, Jane Addams, Charles Francis Adams Jr., two former presidents, and Andrew Carnegie, who offered to buy the Philippines from the U.S. government for $20 million in order to give the country its freedom.

Carnegie was unlike any other robber baron. An avowed atheist whose father had been involved in the radical Chartist movement in Scotland, Carnegie summed up his life's motto in *The Gospel of Wealth* as "The man who dies thus rich dies disgraced."

Carnegie had a lot of giving to do to avoid such ignominy, especially after selling his steel interests to J. P. Morgan in 1901, a move that created the largest company in the world, U.S. Steel. Carnegie's favorite philanthropy was putting up buildings for worthy causes, with the construction of more than 2,800 Carnegie libraries around the world standing as one of the all-time great displays of altruism. Carnegie also financed the Palace of Peace, a grand building in the Netherlands that would house the Permanent Court of Arbitration. This long-cherished goal of the peace movement finally became a reality as a result of the 1899 Hague Convention, called as part of the international pushback against imperialism and war. It was an outcome "far more than any of us dared expect," according to Andrew Dickson White, who served as lead U.S. delegate.

White had been the cofounder of Cornell University and its president for two decades, as well as U.S. ambassador to Germany (twice) and Russia. He was also a historian who authored a book on the hyperinflation of the French Revolution (it was written in defense of the gold standard) and a major peace activist. Among White's beliefs was that Japan and China would become English-speaking, Christian countries in fifty years' time if one thing happened: English were set to a phonetic alphabet.

White was far from alone in his thinking. According to the philologist Francis March, "Complete simplification of our spelling would make our language the easiest to lern [*sic*] of all the great languages, help us in the race for commercial supremacy, and make English the world-speech."

A global tongue was an idea whose time seemed to have come. At the 1849 Peace Congress, men like Cobden had dismissed the idea of a universal language as overly utopian, but toward the end of the nineteenth century the idea had come into vogue, most astonishingly with invented languages.

The first invented language that people took seriously was Volapük. In the 1880s, Volapük boasted hundreds of clubs and several well-attended international conventions, and was even the name of the dog of Frances Folsom, President Cleveland's twenty-one-year-old wife. Volapük was soon displaced, however, by the even more popular Esperanto. First described in the book *International Language* in 1887 by Ludwig Zamenhof under the pseudonym Doktoro Esperanto ("Dr. Hopeful"), its creation was rooted in Zamenhof's belief that a universal language would help lead to world peace. It brought a metric-like simplicity to language, with rules of grammar Zamenhof claimed could be learned in an hour. However popular it became, however, Esperanto was far from being serious competition for French.

It was not pure chauvinism for Victor Hugo to have stated that French would be the universal language of the brotherhood of man. It was so well entrenched as the second language of the world that French text could go untranslated into the U.S. congressional record. But with the spread of the British Empire—and now an American one—English was poised to give French a run for its money. But first, many believed, it had to overcome its one crippling defect.

At the turn of the century, the simplified spelling movement was a quarter of a century old, having started in earnest at the Philadelphia Centennial Exhibition, largely under the leadership of Melvil Dewey. At the time, Dewey paired orthographical and metrological reform, convening studies of educators who "unanimusli" concluded that together the two reforms would save three to four years of education for the average schoolchild, a tremendous amount in a day when most kids went only to grade eight.

The logic and ease that the metric system offered over yards, miles, and pounds was mirrored in what a phonetic alphabet could do for an English orthography that wore the legacy of its words on its proverbial sleeve.

Realize just *how* archaic our spelling is. Pronunciation keys in diction-aries and spelling bees are all but unknown outside of English-speaking countries. The alphabets of other European languages have often been slowly and painstakingly molded under the direction of national acade-mies, exactly the kind of thing the Anglo-Saxon world had proven itself allergic to. And indeed, spelling reform of any kind failed to get very far, being tarred with the stigma of being unrealistic.

Just before century's end came a breakthrough of sorts. The National Education Association approved a list of spelling changes to twelve words as sensible as it was modest, the main aim of which was the elimi-nation of silent letters at the end of words. The new spellings included *program*, *catalog*, *thru*, *altho*, and *thoro*, all of which saw quick adoption in certain dictionaries.

The list was the brainchild of E. O. Vaile, the editor of an education journal, *Intelligence*, and publisher of books on spelling reform dating back to the 1870s with titles like *Our Accursed Spelling: What to Do with It*. Vaile had in mind a far longer list of reformed words and was looking for someone to fund a blue-ribbon commission to certify and promote it. To that end, Vaile contacted Carnegie, known to be sympathetic to the cause. Then Vaile made the mistake of mentioning to Melvil Dewey what he was doing.

No longer the firebrand young reformer, Dewey was now in his fifties and had kicked around a bit. His career had been checkered from the start. Although he had gotten three reform organizations off the ground in 1876, only the ALA and *Library Journal* ever managed a sustainable cash flow, a source Dewey unscrupulously used to support his faltering simplified spelling and metric journals. Even when confronted by those whose pockets he was dipping into, Dewey held himself beyond reproach, on account of the purity of his intentions. His zealousness had always appealed to Frederick Barnard (who helped him out of one of his most serious, near-criminal financial jams), but those who worked for Dewey found him unsympathetic, if not insufferable, and at times a bully, who was not above being manipulative or downright lying to get what he wanted.

That said, Dewey had proven himself an unstoppable force. He continued putting out new editions of his decimal system and headed up the State Library of New York. He also created Lake Placid, a retreat for like-minded Progressives who loved winter sports, hiking, and simplified spelling, which was used on everything from menus to signs for the Adirondack Loj, which has been confusing visitors ever since.

After Vaile told Dewey about his contact with Carnegie, Dewey began to court the philanthropist, with whom he'd already had library-related dealings, while simultaneously telling Vaile he doubted the Scottish-born businessman would help. At a dinner in early 1902, Dewey talked Carnegie's ear off for four hours over simplified spelling; afterward, he let it be known that he'd gotten "the canny Scotsman" to pledge (a perfectly decimal) $10,000 a year for ten years to finance a simplified spelling organization, which Dewey was angling to lead. More than a mere spelling reform group, however, Dewey and others he was recruiting imagined a veritable Academy that would guide the development of the language itself. Dewey set about putting together a board, but before Carnegie would commit the money, he insisted that Dewey gain promises from twenty prominent people to use the dozen NEA-approved spellings in their everyday writing. This attached a new string so far as Dewey was concerned, and he was left with egg on his face with potential board members while still trying to reel in the big fish.

Things got worse for Dewey when he became embroiled in a contro-versy over Jews being excluded from his Lake Placid Club, a common practice in the WASP-ridden Adirondacks, but one that helped lead to his ouster as state librarian after fifteen years on the job. In a letter, he tried to prove his lack of anti-Semitism to Carnegie, largely by suggest-ing the millionaire donate some money to a worthy Jewish cause. With his name in the papers, however, Dewey was hardly the man to be the face of a Simplified Spelling Board that would include the likes of Mark Twain, Andrew Dickson White, Henry Holt, and President Theodore Roosevelt.

CARNEGIE ASSAULTS THE SPELLING BOOK read the *New York Times* headline in March 1906. Backed by the canny Scotsman and his board of heavyweights, simplified spelling was at last taken seriously. The Spelling Board released a list of three hundred words it had pruned according to a set of basic rules. The list included spellings advocated by Webster but never universally adopted—like "color" for "colour"—but also more novel ones, such as the dropping of silent letters to create spellings like *gage* and *fulness*, and advocating the use of *-t* instead of *-ed* where phonetically appropriate, with results like *dropt*, *fixt*, and *lookt*.

The Spelling Board had a measured plan for how it would persuade influential people and organizations to accept its new orthography, a strategy that all but went out the window when one of its members, President Roosevelt, up and issued an executive order directing the Government Printing Office to immediately adopt the list.

"The purpose simply is for the Government, instead of lagging behind popular sentiment, to advance abreast of it, and at the same time abreast of the views of the ablest and most practical educators of our time as well as the most profound scholars." Here was the politician Frederick Barnard had waited for all his life.

In the press, however, Roosevelt was lambasted. The *Baltimore Sun* wondered if the president's order covered his own name: "Will he make it 'Rusevelt' or will he get down to the fact and spell it 'Butt-in-sky'?" Overseas, British reaction verged on the vitriolic, with news headlines calling Roosevelt an anarchist and the *London Evening Standard* stating, "How dare this Roosevelt fellow, the temporary President of an amiable Republic, presume to dictate to us how to spell a language which was ours while America was still a savage and undiscovered country?" But while the reformed orthography was denounced by the likes of Arthur Conan Doyle, certain Britons did support the effort, including the primary editor of the *Oxford English Dictionary*, James Murray, who joined Carnegie's board.

If nothing else, Roosevelt's action heightened interest in the list. School boards around the country said they, too, would adopt the new spellings, and some newspapers put them into everyday practice.

The tide began to turn, however, when Chief Justice Melville W. Fuller refused all documents containing simplified spelling for the business of the Supreme Court. Congressmen tired of being run over roughshod from the bully pulpit saw the whole affair as just one more example of the president's tyrannical leadership style, and delighted in piling on over an issue that was increasingly turning into a humiliation for the hero of San Juan Hill. In December, Congress came out ferociously against Roosevelt's order, and by Christmas, simplified spelling was gone from government publications.

For years, Carnegie kept supporting the movement to the eventual tune of over a quarter of a million dollars, but the lack of progress frustrated a man used to getting his way.

"A more useless body of men never came into association," an exasperated Carnegie wrote to Spelling Board president Henry Holt in 1915, explaining why he was going to cut off funding. "I hav been patient long enuf."

Wartime Measures

I WANT **YOU** FOR THE U.S. ARMY! The image of a goateed Uncle Sam with star-emblazoned top hat became iconic when it started appearing on recruiting posters, part of an unprecedented wave of propaganda America unleashed in support of the war effort. The purchase of Liberty bonds was the most popular subject for posters, followed by those related to food. "Are you a victory canner?" women were asked, while children were implored to eat corn cereal, oatmeal, and hominy so as to leave the nation's wheat for its soldiers. With America's farm boys off in Europe, those on the home front were asked to plant victory gardens. "Get your hoe ready!" declared a 1918 poster that depicted Uncle Sam ordering a little clock-headed soldier to retreat. "Congress passes Daylight Saving Bill" read the caption at top.

The extra hour of garden time was just one of the many benefits of summer hours. The idea of manipulating timepieces to take advantage of unused morning sun went back at least as far as Benjamin Franklin, but it was the London house builder William Willet, an avid equestrian and golfer, who kicked off the daylight saving movement with his 1907 book, *The Waste of Daylight*. His campaign was supported by Arthur Conan Doyle (more of a fan of fiddling with the clock than with spelling, apparently) and a young Winston Churchill, but it was only with total war and its irrefutable logic to economize at all costs—and mobilize the home front—that the reform took off. It was first instituted in Germany and

Austria-Hungary, a bitter irony for those Brits who had spent a decade lobbying for it.

The law adopting daylight saving time for the United States was the Standard Time Act of 1918, which for the first time legally established the nation's time. In so doing, it regularized the railroad-drawn time zones that had informally been put into effect decades earlier, moving such places as Savannah into the Eastern time zone, where its residents no longer had to worry about eating summer dinners in the dark.

By then, Standard Time was well established on the other side of the Atlantic. A movement for all central Europe to go on Greenwich +1 hour had stalled in the late 1880s over Prussian unwillingness to give up Berlin time, but a robust speech in the Reichstag by ninety-one-year-old Helmuth von Moltke quelled such conservatism. The legendary architect of crushing victories over Austria and France, von Moltke argued that a single German time was imperative for war mobilization purposes and national unity; his death the following month underscored his words, reported in papers across the world. By the mid-1890s, nearly all of continental Europe was on Greenwich time, with the notable exception of the French, many of whom couldn't stand the thought of being on *l'heure anglaise*. Even when France finally did get with the program, shortly before the Great War, the switch was to "Paris mean time, retarded 9 minutes 21 seconds," a rather tortured way of saying Greenwich mean time. Or rather, of not saying it.

A third tool in the kit of modern clock time also got a boost from war. The 24-hour clock starting at 0:00 midnight had seen sporadic and limited adoption since Sandford Fleming began agitating for it, being instituted first in Italy by order of King Umberto I in 1893. By the armistice, however, it was in use across Europe, again thanks to the drive for military efficiency.

The American army adopted the 24-hour clock but dropped it once the war was done, as part of the return to normalcy (it would take another world war to make it stick). Ads badgering kids to eat hominy disappeared along with victory gardens, and most farmers wanted daylight saving to follow. Summer hours were despised in rural areas, where it had the effect of farmers *losing* an hour of sunlight, as getting ready for market required them to get up hours before other workers. The only reason to continue it, as far as they saw it, was so city folk could golf. A bill to repeal daylight saving time passed over President Wilson's veto, and most believed that gerrymandering of the clock had ended forever.

Around the same time, a pamphlet entitled *Keep the World War Won* came out, contending that the "secret" of the celebrated efficiency of the German military was the metric system. According to its author, the early lack of a single standard among the Allies had been counted upon by the Kaiser to sow confusion and caused a two-month delay in battle preparedness at a cost to the Allies of two hundred thousand dead and four hundred thousand wounded. To drive the point home, the pamphlet suggested a monument be erected dedicated to those soldiers "who were needlessly killed [for the] lack of world metric."

Propaganda loosely connected to the truth was a hallmark of a new pro-metric group that advocated and agitated under many letterheads— the World Trade Club, the Foreign Trade Club, the All-America Standards Council, and more—but were all funded by the same Albert Herbert, a San Francisco businessman called Mr. Z by his opponents for his prefer- ence to remain anonymous. Herbert's group bombarded Congress with a hundred thousand petitions as well as producing pamphlets and a journal called *The Weekly Metergram*, the greatest hits of which were collected into the five-hundred-plus-page tome *World Metric Standardization: An Urgent Issue*. One of the group's big ideas was to rebrand the metric system by de-Frenchifying it and making its heroes a pair of Scotsmen.

World Metric carried for a frontispiece a photo of the recently deceased Andrew   Carnegie—"Pre-Eminent   Advocate   of   World   Metric Standardization"—and was dedicated to James Watt, who was given credit for inventing the metric system. The latter absurd claim rested purely on Watt's stated desire for decimal measures in general, while the former one was extrapolated from a letter Carnegie had written decades earlier in praise of the meter as part of a committee of manufacturers, which the editor saw fit to rename "the Carnegie Metric Committee." The scheme worked to some extent, however, and people began advocating with a straight face for the "Watt scientific measuring system."

Although Mr. Z certainly livened up the issue, the postwar version of the metric debate otherwise felt like *Groundhog Day*. In Congress, bills were put forth in droves never to be voted upon. Samuel Stratton of the Bureau of Standards created a new organization, the Metric Association, which played the role of the defunct American Metrological Society, even holding its first meeting at Columbia and counting Melvil Dewey among its officers. Stratton's old nemeses, Halsey and Dale, resurfaced as well, with new editions of *The Metric Fallacy* at the ready and plenty of blustery propaganda on their own side, such as a collection of letters in *American Machinist* that ran under the heading "What Real He-Men Think of the Compulsory Metric System."

The end result was the same, too. Any suggestion that the United States was about to lose its manufacturing edge rang more hollow than ever, as the war had proven a bonanza for American goods. The area where *The Metric Fallacy* itself was starting to sound false, however, was its claim that the metric system wasn't really as widespread as advertised. Again, total military mobilization changed things, bringing the meter to the corners of France where it had been resisted. But the metric system was spreading to all-new places as well.

As ever, the meter was buoyed by revolution, and the Great War was nothing so much as a wave of them, which started before the assassination of the Archduke Franz Ferdinand and continued after Versailles. Thousands of years of imperial rule were shown the door with the exit of the Ottoman, Qing, and Romanov dynasties. In their place rose a slew of new republics, many of them would-be Japans.

The rump states of the crumbling Eurasian empires one by one went metric or rededicated themselves to the idea, first with the Republic of China, then in Lenin's Russia, and finally in Turkey. Japan itself, feeling which way the wind was blowing, decided to go metric as well. In places where life was still ordered by the creep of shadows or the call of the muezzin, Western time was adopted wholesale. The introduction of literacy in places where it had never existed was preceded by new phonetic versions of the Latin alphabet, none better than Turkey's, whose leader, Kemal Ataturk, was Japan-like in his keen picking and choosing from the West. Outlawing the fez was the Turkish president's most visible reform, but more crucially he modeled Turkey's civil and penal codes on the Swiss and Italian ones, adopted the Gregorian calendar, and even moved the day of rest from the Islamic Friday to the European Sunday.

The horror of the Great War and the determination that it should never happen again gave the greatest impetus ever to the peace movement. The long-desired Congress of Nations finally saw the light of day, albeit under a different name: the League of Nations.

Idealism reigned in the early days of the League. One of the organization's architects, Lord Cecil, proposed at its first meeting that Esperanto be named one of the League's official languages, alongside French and English, and be taught in the schools of member states, a measure eleven delegates voted for.

A more pressing issue than Esperanto, however, was a tool that was spreading all too fast—the Gregorian calendar. Like the meter and the hour, countries adopted it because it was becoming universal, but unlike either of them, the Western calendar was not remotely neutral, nor even particularly functional. Reformers believed that there were far better options to be had, and some even imagined that tables of days could be done away with altogether.

$$^{13}/_{16}$$

THE GREAT CALENDAR DEBATE

A MERICA UNDERWENT no revolution after Versailles, but massive social changes were nevertheless in store for the country that had about as good a world war as anybody. Progressivism was at its height, with a former university president now the president of the entire nation and reformers getting their way on issues that had roots planted deeply in the previous century. The Eighteenth Amendment and the Volstead Act successfully concluded the long struggle to ban booze with the introduction of Prohibition, while the Nineteenth Amendment gave women the right to vote, fifty years after suffragettes had begun organizing. For pacifists, a long-standing dream was realized with the signing of the Washington Naval Treaty on February 6, 1922, finalizing the first successful disarmament conference in history. The very next day, however, an altogether new issue to the American political scene arrived in Washington: calendar reform.

On February 7, calendar activists gathered for a two-day convention in anticipation of congressional hearings, to be held under the same Andrew Volstead whose name had become indelibly linked with Prohibition. In front of Volstead's Judiciary Committee, witnesses were to testify over an audacious bill that had been introduced to Congress the previous year proposing that America should adopt a thirteenth month. It was to be named after the spring (or vernal) equinox, such that the new U.S. calendar would begin January, February, Vern, March.

The thirteen-month bill was based on a plan put together by Joseph U. Barnes, a Minneapolis insurance executive, Prohibition supporter, and Metric Association member. Five years earlier, on January 1, 1917, Barnes had been struck by the fact that the year was beginning on a Monday. "I sat back in my chair and began to cogitate when it would occur again," Barnes testified to Congress, noting that he straightaway decided the calendar could use some improving. Barnes took out a piece of paper, started jotting some figures, and, about forty-five minutes later, had finished creating a new calendar for the world.

His vision was that every year should have thirteen months of four weeks (see page 223). It was so simple, so perfect, that calendars could be done away with altogether. The best part was how the days of the week would never change in regard to dates. Every year began on a Monday and so did every month—forever. If you were born on a Tuesday, your birthday would always be on a Tuesday. The date of Thanksgiving would be permanently fixed.

How was this possible? The reason that no two consecutive years are the same is that every year contains one more day than 52 weeks, or two more in leap years. The year 1922 had 53 Sundays; 1923 would have 53 Mondays. By removing one day of the year from the week (and all leap days), the calendar would stop cycling on different schedules. As a sweetener for the deal, Barnes proposed that both of the out-of-week days be holidays.

Struck by how good his scheme was, Barnes shared it with friends and soon formed what would become the Liberty Calendar Association with a group of fifty Minneapolis businessmen and professionals.

Barnes eventually discovered—as most reformers with eureka moments do—that his idea was hardly new. His calendar could even be put into a category, that of a fixed or perpetual calendar, and his out-of-week day was called a blank day. What was more, calendar reform had been brewing in Europe for decades.

The first blank day calendar of note also had thirteen months, having been devised in 1849 for Positivism, a philosophical and religious system created by the groundbreaking sociologist Auguste Comte. The Positivist calendar was clearly influenced by the French Republican one, which probably had something to do with Comte himself having been born on 30 Nivôse Year VI (January 19, 1798). Like the Jacobins, Comte eliminated saints from the calendar, but rather than using the days to celebrate agricultural tools and vegetables, he commemorated important historical figures, with great ones such as Shakespeare and Gutenberg getting their own months. (Comte's own birthday would have been Moses 19, the "Theocrats of Japan" day.) Whatever its merits, the Positivist calendar was little used outside Brazil, where Comte's theories took such root that the national flag sports a slogan based on a Positivist motto.

As for the European calendar reform movement proper, it dated to 1884, the year of the Meridian Conference, when a contest for the best calendar plan was held under the aegis of *L'Astronomie* magazine, with a substantial cash prize at stake. The winning plan offered a twelve-month fixed calendar split into identical quarters, each with months of thirty-one, thirty, and thirty days. These four quarters contained thirteen weeks apiece, the inverse of Barnes's and Comte's thirteen months of four weeks each. It was the same math as a deck of cards no matter how you dealt it, with blank days for jokers.

In 1900, slight modifications to the equal-quarters scheme were made by a Geneva-based professor of horology, and the idea started to catch fire. The Swiss plan, as it was now called, didn't do away with the need for a calendar, but it proved more popular in Europe than thirteen-month plans or any of various other schemes. Just as statistical congresses had pushed the meter and franc onto internationalist agendas, the biennial International Congress of the Chambers of Commerce provided the impetus for calendar reform. Longtime supporters of universal measures, free trade, and the codification movement, chambers of commerce congresses took up the calendar cause in 1908, and soon activists from countries all over Europe—Britain, Belgium, Italy, and Spain included—were pitching variations of the Swiss plan. At their 1914 meeting in Brussels, plans were made to hold an international calendar conference in Switzerland that fall. The timing couldn't have been worse, but the movement picked up right where it left off after the war, with European leaders in business, religion, and astronomy meeting to discuss the issue.

Barnes hoped that the American calendar conference would raise the international profile of his thirteen-month scheme, but a man arriving in Washington from three thousand miles across the North American continent was about to steal Barnes's thunder. For over a quarter of a century he had been trying to sell his calendar plan to anyone willing to listen, and he was about to find his audience.

Moses, the Greatest Calendar Reformer of All

Moses Cotsworth had been obsessed by one thing his entire life: time. Orphaned at the age of two, Moses was raised by his grandparents and great-grandparents in a rural corner of England near York. His grandfather was a shepherd and his great-grandfather a farmer who had been so poor that he always had lived in sod houses and never been able to afford a watch. On walks through the countryside, the old man amazed young Moses with his ability to tell time by the length and direction of the shadows that fell around them.

Brought up in a house with sundials and shadow pins, Cotsworth was able to recognize that the local Stonehenge-like megaliths called the Devil's Arrows were oversize versions of the same thing. As an adult, Cotsworth became convinced that the pyramid at Giza had also been built as a massive time-telling device—in its case, to keep track of the year—an idea he found support for in *Our Inheritance in the Great Pyramid*. Thrilled to learn that its author was not only still alive, but had a house just the other side of the Devil's Arrows, Cotsworth arranged a meeting.

Charles Piazzi Smyth still wore his Egyptian tarboosh, but he had lost his beloved wife and was not in good health, and under medical orders Cotsworth was allowed only fifteen minutes with him. The old astronomer mumbled and was hard to understand, but on the crucial question—Was the shortest shadow at Giza cast on the equinox?—Piazzi Smyth gave an emphatic "Yes!" and Cotsworth left convinced that he was correct about the calendrical function of the Great Pyramid.

Cotsworth traveled the world seeking alternative calendar systems, such as the five-bundle stick system used by Native Americans in northwest Canada to keep count of a "year" of five months. At his day job, Cotsworth worked as a statistician for the railroad, publishing what would become the standard work on British railway rates as well as a series of mathematical aids called *Cotsworth's Card Calculators*, which included entries on converting to and from the meter, liter, kilogram, and various currencies. He found his obsession and profession colliding often.

When it came to calculating costs and revenue, the awkward coexistence of the fixed week and the inconsistent month caused constant conflict. Whereas weeks were the reliable time units of business, months skewed statistics horribly, being made up partly of full weeks (three or four) and partly of broken weeks (one or two). Only once every nine Februaries or so did you get a truly perfect month of four intact weeks starting on a Sunday. Thirteen of these, Cotsworth believed, would make the ideal calendar.

To promote his plan, Cotsworth collected his research and sank his life's savings into the lavishly illustrated *The Rational Almanac*, published in 1902. With an embossed gold-on-black cover, unusual vertical profile, and nearly five hundred pages, it made an impressive statement. But in terms of creating a groundswell of support for his calendar, *The Rational Almanac* was a dud, released just after what would prove in Europe to be the far more popular Swiss plan.

In 1907, Cotsworth was hired by the government of British Columbia to reorganize its civil service. He found in what would become his adopted country the ideal supporter, Sandford Fleming, of whom Cotsworth would later pen a short biography entitled *The Greatest Canadian*. Fleming had been collecting and presenting European calendar reform plans to the Royal Society of Canada, but in Cotsworth's scheme Fleming spotted something akin to what he had so long fought for—a system of time reckoning for the world that was both neutral and rational. To complement Standard Time, Cotsworth had come up with the Standard Month. With Fleming as its honorary president, Cotsworth set up the International Fixed Calendar League, and with an improved knack for propaganda began to write books like *Time to Fix the Year*. Meeting with acclaim in Canada, Cotsworth seized on Barnes's conference to try to make an impact in the United States.

The calendar convention arranged by Barnes's Liberty Calendar Association opened on February 7, 1922, in the auditorium of the U.S. National Museum of the Smithsonian, which is today the Museum of Natural History. Eleven calendars were examined, including a few nutty ones such as the "first Scottish plan," which contained four- and five-week months and a "metric calendar plan" of ten months that alternated between having thirty-six and thirty-seven days. The most serious contenders were the Swiss plan, presented and advocated by A. F. Beal of the Bureau of Standards, and three thirteen-month schemes, which included Barnes's Liberty Calendar, the assumed winner. But in the end it was Cotsworth's plan that dazzled the participants in the conference. Or rather, it was Cotsworth himself.

In truth, his International Fixed Calendar differed little from the Liberty one. Rather than putting it after February, Cotsworth inserted the extra month between June and July and called it Sol, in honor of the solstice that would take place during it. He also started each month on a Sunday instead of a Monday. This made for a triskaidekaphobe's nightmare, with thirteen Friday the Thirteenths every year, but this was a calendar for the rational, not the superstitious.

More than having just thought out the details better, Cotsworth amazed everyone with his knowledge. No one knew more about calendars than the Quaker from York, not just their history but how the Gregorian one presented a roadblock to progress and efficiency. He perfectly articulated the superiority of the thirteen-month calendar, stressing how its regularity would unify time by allowing the addition of hands for the month and day to timepieces, as ones for the minute and second had been added with the invention of the pendulum clock.

The convention passed unanimous resolutions both in favor of Cotsworth's plan and to express thanks for his "enlightening and practical presentation." (They were introduced by the Minneapolitan Barnes, showing him to be far from a sore loser.) This moment of triumph was followed the next day by Cotsworth's appearing alongside Barnes and others before Volstead's House Judiciary Committee.

In testifying, reformers hit hard upon the practical problems of the existing calendar. As an example, Cotsworth noted that that year's February would have 14 percent less earning time than the months abutting it. Witnesses also tried to show how a thirteen-month year was less radical than it seemed, pointing out that a form of it was already used by many companies, not to mention the two-thirds of humanity that still lived by moons, of which more than twelve occur every year.

That the overwhelming preponderance of the world's population was not yet using the Gregorian calendar spoke to the urgent need to implement a universal one. Cotsworth described the remarkable variety of calendars that coexisted, with five in use in both Cairo and Constantinople. The congressman who had introduced the calendar reform bill spoke of the Gregorian as "this cumbersome calendar of the ancients," unfit for an age of telephones, wireless telegraphy, and airplanes. Witnesses also kept noting how relatively recently the Gregorian reform had been adopted, when George Washington was already an adult.

As in the rest of the British world, Wednesday, September 2, 1752, was followed in colonial America by Thursday, September 14. The loss of those eleven days was the shock therapy needed to catch up from the overly leap-yeared and thus slow Julian calendar and caused the birthday of Washington to be moved from February 11 to February 22. What was more, his birth *year* changed from 1731 to 1732, as New Year's Day was moved from March 25 to January 1. All of this conspired to make 1752 a mere 293 days long, demonstrating that radical calendar reform was not unprecedented.

The best point, however, was the one made most repeatedly: if the thirteen-month calendar were adopted, no sane person of the future would ever propose returning to the present one.

The Calendar of the Future Is Now

Although supporters of a thirteen-month plan dominated the calendar reform hearing, there was a cordial but sharp dissent issued from A. F. Beal. An expert on time matters working at the Bureau of Standards, Beal maintained that most of the nation's scientists were against the thirteen-month calendar scheme, and cited support among astronomers for the twelve-month Swiss plan that divided the year into equal quarters. A strong believer in the necessity of reform, Beal further maintained that the less radical Swiss plan was the only one with a realistic shot of being universally adopted, noting that the plan was favored in Europe, where the matter was much farther along.

With the war over, the International Chamber of Commerce (as the organization that succeeded the old congresses was called) picked up where it had left off and organized a commission to study calendar reform. A similar committee had also been put together by the International Astronomical Union and was composed not only of scientists but religious officials, including Cardinal Mercier, the Belgian war hero, who would preside. The reason for the involvement of religious figures was the recognition from most everyone that if calendar reform changed nothing else, it should fix Easter.

That the most important holiday on the Christian calendar wandered across thirty-five days of March and April like a drunken sailor was both one of the best things calendar reformers had going for them and their biggest obstacle. Easter falls on the first Sunday after the first full moon on or after the spring equinox, which is complicated enough, but by ignoring astronomical observation in favor of its own particular method for setting the equinox and full moons, the Catholic Church had made its calculation so cumbersome that it took an entire mathematical discipline called *computus* to figure out when to celebrate the Resurrection.

The question of the blank day was also potentially religious, as the seven-day week had its basis in the story of Genesis. Most of the major Christian churches, however, showed a willingness to accept blank days, and so long as the calculation of movable feasts was left to them, they presented themselves as no obstacle to calendar reform per se.

Both the committees of the International Chamber of Commerce and the International Astronomical Union favored a fixed calendar reform, and they appealed to the League of Nations to make it happen. It was the sort of issue the Geneva-based organization had been formed to handle, and some believed that the creation of a new international calendar could prove to be a defining moment.

The League of Nations took up the question in 1923 and formed its own calendar reform committee the year following. By this point, the International Fixed Calendar League was claiming offices in London and Washington as well as Vancouver. Cotsworth had been traveling the world for two years in search of converts, and nearly bankrupted himself in the process. In Washington, he had won over Charles Marvin, the chief of the Weather Bureau, who had worked under Cleveland Abbe during the push to spread Standard Time globally. Marvin shared Abbe's views on how science could benefit from a more rational approach to time, and would dedicate himself wholeheartedly to the new cause.

But no convert would prove so important as George Eastman. The inventor of the first easy-to-use camera and founder of Kodak, Eastman was not only one of America's foremost businessmen but among its greatest philanthropists. Unlike charitable robber barons such as John D. Rockefeller and the recently deceased Carnegie, however, Eastman gave quietly. He eschewed putting his name on buildings, for instance, and donated close to $20 million to MIT as "Mr. Smith."

Eastman could easily have seen Cotsworth as a crank when the two met. Certainly, the thirteen-month idea had been given a hard time by some in the press, who shuddered at the idea of the fourth of July coming on the seventeenth of Sol or one out of thirteen people losing their birthdays (i.e., anyone born on the 29th, 30th, or 31st of any month). As a businessman, Eastman was intrigued by the economic benefits a new calendar could offer, but as a human being the millionaire was moved by the personal tale of Cotsworth and how he had left his high-paying work and given all of his money and time to a cause aimed only at the betterment of the world. Eastman believed that the Quaker Cotsworth embodied all the positive attributes of the Society of Friends.

The millionaire would prove to be Cotsworth's fairy godfather. In what one biographer termed "a staggering effort," Eastman set the Quaker from York up with a New York office and put his best people from Kodak at Cotsworth's disposal, including Marion Folsom, the future architect of Social Security, as well as the company's public relations director and an international lobbyist who traveled the world currying favor for the plan. But even better than money and personnel was the reputation and Rolodex of Eastman himself.

Almost overnight, the thirteen-month plan went from being just some fool idea to one that needed to be taken seriously, as did Cotsworth himself. Eastman was a less tricky benefactor than Carnegie had been for simplified spellers, putting himself fully behind the International Fixed Calendar, to the point of lending his name to articles and a book that laid out the argument for its adoption.

Having considered 195 reform schemes from 54 countries, the League of Nations committee rendered its findings in 1926, showing favor to the

Cotsworth scheme and the Swiss equal-quarters plan. As the next step in the process, the League asked countries to put the matter to national committees, and here Eastman's contacts proved invaluable.

Having discussed the matter with Secretary of State Frank Kellogg, Eastman assumed the task of organizing the National Committee on Calendar Simplification. With the assistance of Charles Marvin, Eastman packed the committee with powerful national leaders, including Henry Ford, the secretary of labor, the publisher of the *New York Times*, and the old metric advocate and former Bureau of Standards chief Samuel Stratton. In its 1929 report, the committee came out strongly for the thirteen-month plan, which further won the endorsement of corporations like U.S. Steel and General Motors as well as the National Academy of Sciences, which voted for it unanimously. Momentum reached the point where the press began reporting on the coming calendar change as inevitable, and a joint resolution was sought in Congress for hosting an international calendar simplification conference.

Emerging religious opposition, however, threatened to torpedo the whole thing. It came from Orthodox Jewish leaders and Seventh Day Adventists vehemently opposed to any blank day calendar. As can be deduced from their name, Seventh Day Adventists took the days of the week seriously, with a central tenet of their faith being Saturday observance of the Sabbath. In an age of six-day workweeks, this was a problem, particularly when the imposition of late-nineteenth-century blue laws made working on Sunday a crime in certain places, causing some Adventists to go to jail for keeping a different schedule. Both Adventists and Orthodox Jews insisted they would continue to observe the Sabbath every seven days no matter what the civil calendar said, believing that the week had been kept in uninterrupted succession since creation, which literal readers of the Bible placed somewhere between 4004 and 3761 B.C.

Leading the charge inside Congress against the joint resolution was

Rep. Sol "the Music Man" Bloom, whose claim to fame was composing the belly-dancing tune that goes with "There's a place in France where the ladies wear no pants." Bloom made a June 1929 speech in Congress claiming that calendar reform stomped on freedom of religion. Meanwhile, the Adventist press compared the 13-month calendar to the atheistic French Revolutionary one and what was happening in the Soviet Union, where a new five-day week similarly sought to obliterate the Sabbath habit.

Seeking to defuse religious opposition, Cotsworth cowrote a book called *Moses, The Greatest Calendar Reformer of All*. (The title referred to the Old Testament prophet, not the author.) The book outlined how scientific discoveries such as evolution and carbon dating had rendered creationist timelines absurd, and pointed out that consistent observance of the week hadn't begun until well into recorded history besides. The experience of Magellan's crew discovering they were keeping the wrong Sabbath was also brought up. Did their loss of a day mean that they were no longer in God's grace? A blank day was little different from crossing the international date line.

Although the bill to host an international conference got squelched, Cotsworth's movement continued to gain momentum, with corporations switching to thirteen-period accounting in droves. With plenty of reason for optimism, he and his associates turned their sights to an upcoming League of Nations meeting that would take up calendar reform as its featured topic. After so many years and so much struggle, it looked as if it might finally be Cotsworth's moment. Not only would his calendar plan be front and center in Geneva, but the Quaker from York was named a special "Expert" to the League's Committee on Calendar Reform. One newspaper wrote that "Cotsworth" was the name on the lips in every foreign capital.

But it was now Moses Cotsworth's turn to be upstaged, as a new player entered the fray via an old friend of reform.

The World Calendar

Although she was almost fifty, Elisabeth Achelis was still looking for her life's work. Passionately Protestant and Progressive, Achelis was a minor heiress who had long resisted what was expected of her, having served in the Red Cross during the Great War and never marrying.

In the late summer of 1929, Achelis traveled from her New York City home to an upstate retreat beloved of Progressive folk. The Adirondak Loj had hiking trails, while the Watersyd Boathouse boasted a menu that would offer items like pe sup, hucklberi pi, stud prunes, and cofi. Here she listened to a lecture entitled "On Simplifying Life," delivered by the founder of Lake Placid, Melvil Dewey.

Better than fifty years removed from the heady days of 1876, old Dewey had done well in the end. At Lake Placid, he had created an atmosphere where people lived according to his rules, most of them gladly so. His son Godfrey was following in his footsteps, having become a driving force in the simplified spelling movement as well as in winter sports, having just engineered the winning bid for Lake Placid to hold the second Winter Olympics.

In his talk, Dewey hit upon three of his causes, two old ones—the metric system and simplified spelling—and a new one, the thirteen-month calendar, which greatly appealed to his sense of economy and rational purity. Achelis, on the other hand, found herself agitated by the scheme, a feeling she was unable to shake even once she returned home to Manhattan. The division of the year by the number thirteen was irksome, and the whole plan seemed too radical and unbalanced to her. Shortly after, the *New York Times* published a letter presenting a European alternative to the thirteen-month scheme that was all the rage in America—the twelve-month, equal-quarters Swiss calendar. Reading its details, Achelis knew she had found her cause.

Fueled by her formidable brains, ambition, and bank account, Achelis put herself at the forefront of the calendar reform movement not only in America but globally. With remarkable speed, she formed an advocacy organization, sponsored research into public opinion, started a journal, and published her beliefs, rebranding the fading Swiss plan as the World Calendar and calling the annual year-beginning blank day Worldsday. While an international holiday of peace had been part and parcel of nearly all fixed calendar schemes, Achelis hit the point hard, saying that universal observance of Worldsday would draw peoples and nations closer together. Previously, calendars had all been religious or national creations, formed to mark and celebrate events for a given people and therefore exclusionary. A common, neutral calendar was a step toward binding the world into a single community.

Achelis quickly garnered support. She early on set to swaying Mohandas Gandhi, whom she met in his rented house in London, where he was participating in negotiations on the fate of the Raj with the British government. Calendars were a particular problem in India, where at least thirty were in use. The World Calendar won with Gandhi's approval because it was a 12-month plan, it was rational, and it was free of nationalist and religious taint. He compared it to Esperanto and universal coinage, two other causes he supported.

Even those Achelis didn't win over respected her. A month before he died, Dewey became a member of her organization, writing, "I have from the 1st been loyal to the 13 month skeme but hav an open mind & wil be glad to put the JOURNAL in our librari wher our members can get yur syd of the question."

Although he and Achelis treated each other respectfully, her janie-come-lately entry had to come as a crushing blow to Cotsworth, whose thirty-five-year quest had seemed on the cusp of success.

Both spoke to the League preparatory committee that met in June 1931, largely to attack each other's calendars. By this point, the League had looked at more than 500 schemes in total, which they had boiled down to 52 basic plans, but it was clear that the two American-supported ones were the only with any chance of success. Of the national committees that reported, 13 of 16 wanted a fixed calendar, and most of them preferred Cotsworth's.

The 13-Months Calendar

Proposed to be Perpetual and Universal *from 1934.*

The League of Nations has decided to invite all Nations to simplify
the Calendar, by International Conference during October, 1931.

The 3-Methods of Reform to be then considered, are printed on the back.

EVERY MONTH—4 WEEKS.

Sun	Mon	Tue	Wed	Thu	Fri	Sat
1	2	3	4	5	6	7
8	9	10	11	12	13	14
15	16	17	18	19	20	21
22	23	24	25	26	27	28*

Every week-day on its 4 Fixed Monthly Dates. *Leap-day, June 29 and *Year-day, Dec. 29,
—both to be International Holidays, after Saturday. Christmas and Holidays on Mondays.
The new 28-day month "Sol" would gather up all 29th, 30th and 31st
dates, by absorbing the last 13 days of June and the first 15 days of July.
All Church Festivals would be "Fixed" to perpetual dates, by fixing
Easter Sunday on April 15th (New Style), and Whit-Sunday on June 8th.

Advantages of the proposed "Fixed Calendar" are summarized on the back.

Every Month like February in 1931

1931 - CALENDAR - 1931

JANUARY	FEBRUARY	MARCH	APRIL

SUN	MON	TUE	WED	THU	FRI	SAT
				1	2	3
4	5	6	7	8	9	10
11	12	13	14	15	16	17
18	19	20	21	22	23	24
25	26	27	28	29	30	31

SUN	MON	TUE	WED	THU	FRI	SAT
1	2	3	4	5	6	7
8	9	10	11	12	13	14
15	16	17	18	19	20	21
22	23	24	25	26	27	28

SUN	MON	TUE	WED	THU	FRI	SAT
1	2	3	4	5	6	7
8	9	10	11	12	13	14
15	16	17	18	19	20	21
22	23	24	25	26	27	28
29	30	31				

SUN	MON	TUE	WED	THU	FRI	SAT
			1	2	3	4
5	6	7	8	9	10	11
12	13	14	15	16	17	18
19	20	21	22	23	24	25
26	27	28	29	30		

MAY	JUNE	JULY	AUGUST

SUN	MON	TUE	WED	THU	FRI	SAT
					1	2
3	4	5	6	7	8	9
10	11	12	13	14	15	16
17	18	19	20	21	22	23
24/31	25	26	27	28	29	30

SUN	MON	TUE	WED	THU	FRI	SAT
	1	2	3	4	5	6
7	8	9	10	11	12	13
14	15	16	17	18	19	20
21	22	23	24	25	26	27
28	29	30				

SUN	MON	TUE	WED	THU	FRI	SAT
			1	2	3	4
5	6	7	8	9	10	11
12	13	14	15	16	17	18
19	20	21	22	23	24	25
26	27	28	29	30	31	

SUN	MON	TUE	WED	THU	FRI	SAT
						1
2	3	4	5	6	7	8
9	10	11	12	13	14	15
16	17	18	19	20	21	22
23/30	24/31	25	26	27	28	29

SEPTEMBER	OCTOBER	NOVEMBER	DECEMBER

SUN	MON	TUE	WED	THU	FRI	SAT
		1	2	3	4	5
6	7	8	9	10	11	12
13	14	15	16	17	18	19
20	21	22	23	24	25	26
27	28	29	30			

SUN	MON	TUE	WED	THU	FRI	SAT
				1	2	3
4	5	6	7	8	9	10
11	12	13	14	15	16	17
18	19	20	21	22	23	24
25	26	27	28	29	30	31

SUN	MON	TUE	WED	THU	FRI	SAT
1	2	3	4	5	6	7
8	9	10	11	12	13	14
15	16	17	18	19	20	21
22	23	24	25	26	27	28
29	30					

SUN	MON	TUE	WED	THU	FRI	SAT
		1	2	3	4	5
6	7	8	9	10	11	12
13	14	15	16	17	18	19
20	21	22	23	24	25	26
27	28	29	30	31		

May and August have days in 6-weeks—Nine other months have days in 5-weeks.
Coming dates for Easter: 1931, April 5; 1932, March 27; 1933, April 16.

From the International "Fixed Calendar" League, 1, Regent St., London.

THE WORLD CALENDAR

JANUARY	FEBRUARY	MARCH
S M T W T F S	S M T W T F S	S M T W T F S
1 2 3 4 5 6 7	1 2 3 4	1 2
8 9 10 11 12 13 14	5 6 7 8 9 10 11	3 4 5 6 7 8 9
15 16 17 18 19 20 21	12 13 14 15 16 17 18	10 11 12 13 14 15 16
22 23 24 25 26 27 28	19 20 21 22 23 24 25	17 18 19 20 21 22 23
29 30 31	26 27 28 29 30	24 25 26 27 28 29 30

APRIL	MAY	JUNE
S M T W T F S	S M T W T F S	S M T W T F S
1 2 3 4 5 6 7	1 2 3 4	1 2
8 9 10 11 12 13 14	5 6 7 8 9 10 11	3 4 5 6 7 8 9
15 16 17 18 19 20 21	12 13 14 15 16 17 18	10 11 12 13 14 15 16
22 23 24 25 26 27 28	19 20 21 22 23 24 25	17 18 19 20 21 22 23
29 30 31	26 27 28 29 30	24 25 26 27 28 29 30 [W]

JULY	AUGUST	SEPTEMBER
S M T W T F S	S M T W T F S	S M T W T F S
1 2 3 4 5 6 7	1 2 3 4	1 2
8 9 10 11 12 13 14	5 6 7 8 9 10 11	3 4 5 6 7 8 9
15 16 17 18 19 20 21	12 13 14 15 16 17 18	10 11 12 13 14 15 16
22 23 24 25 26 27 28	19 20 21 22 23 24 25	17 18 19 20 21 22 23
29 30 31	26 27 28 29 30	24 25 26 27 28 29 30

OCTOBER	NOVEMBER	DECEMBER
S M T W T F S	S M T W T F S	S M T W T F S
1 2 3 4 5 6 7	1 2 3 4	1 2
8 9 10 11 12 13 14	5 6 7 8 9 10 11	3 4 5 6 7 8 9
15 16 17 18 19 20 21	12 13 14 15 16 17 18	10 11 12 13 14 15 16
22 23 24 25 26 27 28	19 20 21 22 23 24 25	17 18 19 20 21 22 23
29 30 31	26 27 28 29 30	24 25 26 27 28 29 30 [W]

W—World Day. December W (365th day), a world
holiday, follows December 30th every year. Leap-
year Day, June W, another world holiday, follows
June 30th in leap years.
Easter Sunday, April 8th.

Both the International "Fixed Calendar" League and the World Calendar Association published a lot of material promoting their reforms in anticipation of the 1931 League of Nations conference. Notice that the annual blank day under the thirteen-month plan would be December 29, while in the World Calendar plan it would be December W. The Fixed Calendar League also proposed celebrating all holidays on Mondays, the one reform that actually succeeded with Congress passing the Uniform Monday Holiday Act of 1968.

Even though calendar reform was the marquee issue at the League's Conference on Communications and Transit that fall, nothing was decided. Only four of forty-one nations voted for plans—two for each of the fixed schemes—while the rest were split between wanting less reform, no reform, or not being sure what they wanted. Clearly some governments had been spooked by religious opposition, no matter how relatively marginal.

Far from being a signature achievement, calendar reform wound up being a precursor to all that was about to go wrong with the League, which up to then had been a relative success. It also showed how hard it would be for measurement reform of any kind to happen in the future. Even in the days when the choice came down to the British way or the French way, reform was hard enough—what chance did change have at a conference of forty-plus equal partners?

The biggest problem was that both of the contending calendars were worthy and neither clearly better, making the choice between them nearly impossible. The true shame was that a choice needn't have been made. The Cotsworth and Achelis calendars were in agreement over the basic structure of a fixed calendar: an even 52-week year starting on a Sunday, with one blank-day holiday introducing each year and another at the midpoint of leap years. The arrangement of the months was window dressing and could have been left to individual countries—and bookkeepers—to decide for themselves. If Achelis and Cotsworth had gotten together and fought for a basic fixed calendar to replace the Gregorian, the fixed calendar would have won, and the days of the week would have stopped their eternal cycling through the years.

Or not.

The 1931 League discussion on the calendar might just have come too late. Any hope that the 1929 Wall Street stock market crash was a blip had proven illusory, and the idea of a global economic dark age was becoming a reality. Even had a fixed calendar found overwhelming

League support, would it really have been instituted, considering all the turmoil about to come unleashed?

With the suicide of a physically deteriorating George Eastman in 1932, the thirteen-month scheme lost its most powerful backer and retreated, even if Cotsworth continued to soldier on. Achelis managed to gain the support of a few South American nations and to keep the World Calendar on the League agenda, but events overtook her and her work.

Although no one could know it at the time, the era of the revolutionary reforms that changed the way the world was measured had come to an end. The question of reform was now one of adoption. For better and for worse, the basic measures of the world were set.

With one exception. The most venerable of all measures was getting transformed in such a way that it would cease to remain a traditional measure at all.

14/16

or Seven Eighths

SHOCKS TO THE SYSTEM

WHILE THE Great War had boosted several universalist projects, one of the great liberal successes of the nineteenth century faltered in the conflict's aftershocks—the international gold standard. As always, nations had turned to paper money in wartime, and this time Humpty Dumpty couldn't be put back together again.

War reparations imposed on Germany helped to ruin the economy of the Weimar Republic, causing a hyperinflation that outdid even that of the French Revolution, the previous benchmark, and led to the archetypal story of Germans carting around piles of banknotes in wheelbarrows just to buy bread, which cost upwards of half a trillion marks in 1923. Similar scenes occurred in Lenin's Russia, but these were at least partly by design, as the czar's printing presses were turned on full steam in order to make money worthless and wipe out the cash wealth of the rich, which some Bolsheviks saw as the start of wiping out the idea of wealth altogether.

Even among the neutral states and supposed winners there was currency pain aplenty. Both the Latin and Scandinavian Monetary Unions fell apart during the war, with individual countries getting back on the gold standard only at steep devaluations. Britain spent nearly seven wrenching years trying to put the pound sterling back on gold at its prewar price, and only managed to succeed partially. Some thought it wasn't worth the effort, with John Maynard Keynes writing of the gold standard as a "barbarous relic" in 1924.

The United States, on the other hand, didn't have its gold crisis until the opening days of the new Roosevelt administration. Inaugurated in the middle of a bank run that caused gold deposits to sink to dangerous lows, the new president took a series of unprecedented steps that "nationalized" gold. First, Roosevelt declared that dollars could no longer be redeemed for precious metal, except by foreign banks and other nations, putting the United States on a limited gold standard. His next step was more extreme. At the beginning of 1934, Roosevelt ordered that all the nation's gold be confiscated.

The possession of precious metal became a crime punishable by prison, with some small exemptions made for personal effects such as jewelry. Officially, this was to end "hoarding," but once the government had paid those who had turned over their gold at its traditional price of $20.67 per ounce, Roosevelt devalued the dollar, declaring it would now take $35 to buy an ounce of gold.

All at once, every dollar was worth 40 percent less than it was before. The profit realized by this move—$2 billion, according to the president—went to the U.S. Treasury. It was exactly the sort of thing a universal currency was to have helped prevent, especially one denominated purely by weight. If the United States was using money that defined its value in grams of gold, it would have been harder to debase than a dollar linked to gold by an obscure ratio.

Whether this would have been a desirable situation was a whole other question; Keynes and a rising number of other economists thought not. What's more, it was becoming increasingly clear by 1934 that universalism in general was losing its appeal.

That most powerful of liberal bonds between nations—free trade—began to come unraveled in 1930 with the erection of protectionist tariffs by the United States, an act that worsened the Depression and led to a deteriorating relationship with Japan, which in turn flouted agreements made at the Washington disarmament conference and invaded Manchuria. League criticism led to Japan's departure from the organization, exposing the institution's impotence, which was further proved by its inability to do anything about the invasion of Ethiopia by Fascist Italy and Stalin's annexation of Finland, let alone Hitler's invasion of Czechoslovakia.

In spheres of isolation, nationalism fed on itself, and countries halted all sorts of universalist plans to return to their cultural roots, at times in the service of imperialism. The USSR started a Russification program across Central Asia and Slavic Europe that included replacing the Latin alphabet with the Cyrillic one, while the Chinese forsook phonetic alphabets altogether to return to their traditional script. China also put the brakes on metrication, as did Japan. Metric interest in America, which had been ongoing since the start of the Civil War, ceased to exist.

As for the Second World War itself, the conflict exposed the fault lines between where the meter had been accepted and where it had not. Hitler's domination of continental Europe put him in control of the core metric countries, while the bulk of the fight against the Axis was waged by foot-pound nations. What arms and aid the French Resistance and other Allies received came sized in Anglo-Saxon measures, such that the scientists at the International Bureau of Weights and Measures at Sèvres worried they were seeing the permanent enshrinement of a dual-measurement world, or worse.

Whether or not men like Richard Cobden had been correct in maintaining that international ties would put an end to war, its reverse had proven true—that their unraveling would lead to global conflict. Once victory came into view, the Allies started to plan how to stitch the world back together in such a way that it could never again unwind so disastrously. In July 1944, more than seven hundred delegates of the world's forty-four allied nations gathered at the grand old Mount Washington Hotel in Bretton Woods, an area located in northern New Hampshire. Over three weeks of meetings, the financial strategy of the postwar era was mapped out. The Bretton Woods conference led directly to the

International Monetary Fund and what later became the World Bank, as well as setting the stage for the free-trade agreements of the GATT and negotiations on a more effective successor to replace the moribund League of Nations.

What came to be called the Bretton Woods system, however, had to do with how countries would return to the gold standard. Essentially, they wouldn't. Instead, the currencies of the world would be pegged to the only one still backed by gold. In effect, there was finally a universal currency, and it was the U.S. dollar.

With the world on a dollar standard, English becoming the new language of international affairs, the United States accounting for a third of global GDP, and the seat of world government about to be erected on the banks of New York's East River, the scientists at Sèvres could be forgiven for worrying that the pound and foot were on the verge of taking over the world.

Fear of a Metric Future

"'E could 'a drawed me off a pint," grumbled the old man as he settled down behind a glass. "A 'alf litre ain't enough. It don't satisfy. And a 'ole litre's too much. It starts my bladder running."

Written in the immediate aftermath of World War II, George Orwell's *1984* imagined that the metric system had been imposed on America and Britain by the totalitarian regime of Big Brother. The system's nomenclature fit in perfectly with the structure of Newspeak, Orwell's satire of attempts to invent a rational version of English to serve as an international auxiliary language. The association of the meter with authoritarianism was not entirely unfair, as it could still be said when the novel was published in 1949 that no democracy had ever adopted the meter, and that no people had willingly accepted it. But not for much longer.

As World War I had seen the disintegration of ancient empires, World War II saw the dissolution of modern ones. In India, Jawaharlal Nehru, who took particular pleasure in spurning all things British, quickly broke out the meter sticks and centigrade thermometers as well as a new decimal coinage. He also embraced the World Calendar, finding it superior to the inconvenient and "unscientific" Gregorian one imposed on India by the British for civil purposes. Elisabeth Achelis had never stopped trumpeting the reform, and she found in Nehru a sponsor at the United Nations to stand alongside the trio of Latin American countries she had already recruited. The measure to adopt the World Calendar died in 1955, however, when the United States declared itself unconditionally opposed to any change. This position was delivered by the American ambassador to the UN, Henry Cabot Lodge Jr., who had said of the new world body, "This organization is created to prevent you from going to hell. It isn't created to take you to heaven." Victor Hugo's "Humanity" this was not.

South of the Sahara, colonialism's end reached its critical point in 1960's "Year of Africa," during which seventeen new nations came into existence. As most were former French or Belgian possessions, the metric ranks swelled automatically, but they were joined by former British colonies who like India preferred to go metric rather than remain imperial. In the Far East, Japan, South Korea, and China had all committed themselves to the meter, while in Europe, Greece finally issued a metric order that stuck in 1959—123 years after it had become first non-Francophone nation to officially adopt the meter. The idea that Anglo-Saxon measures would displace the meter had proved as ephemeral as the United States' maintaining 35 percent of the world's GDP. In fact, America itself would begin to consider metric measures again, as a result of the increasing anxiety it was feeling over the internationalist project it could least abide—Communism.

The Red Scare had intensified once it became known in 1949 that the USSR had developed the atomic bomb. The spectacle of McCarthyism showed how deep the fear ran, as did the hysterical American reaction to the launch of Sputnik in 1957. That the USSR had beaten the USA to putting the first man-made satellite into space was bad enough, but the idea that the Soviets were one step away from being able to drop a nuclear bomb from the stars was downright terrifying. The United States became embroiled in the question of how such a thing could have happened, and overnight American confidence in itself began to evaporate. In a way it never had before, science shot to the top of the national agenda.

In 1958, Congress passed the National Defense Education Act, which infused piles of federal cash into the nation's schools and led to an overhaul of science curricula. To fight the perceived superiority of Soviet engineers, the New Math was installed in grammar schools, darkening the spirits of a generation of students. And national legislators debated the meter for the first time in three decades.

"What is this meet-ric system?" the chairman of the House Rules Committee was reported to have asked. His question came in response to a proposal for a three-year study into metrication and was a perfectly understandable reaction, considering how obscure a subject the meter had become to most Americans.

The great line of measurement reformers that had run from Jefferson through John Quincy Adams, Alexander Dallas Bache, Frederick Barnard, Melvil Dewey, and Samuel Stratton had for all intents and purposes been extinguished. The sole present-day link to the past was the Metric Association, but just barely. It was presided over from the 1930s through the '50s by a Dr. John T. Johnson, whose chief accomplishment was a pro-metric textbook he put together for the National Council of Teachers of Mathematics in 1948. It was only with Sputnik and the introduction of bills to study the meter in 1959 that the Metric Association began to revive, and even then it would never attract the talent of the original, let alone that of Frederick Barnard's American Metrological Society.

While America debated whether it was worth considering, the metric system was in the midst of an overhaul that saw even its name change—to SI, for the Système Internationale of units. This took place in 1960, capping a dozen years of changes by the International Bureau of Weights and Measures, the foremost one being the replacement of the meter itself. This time the standard wasn't replaced by a new bar of metal or any physical object at all, but by a definition, which was based on the wavelength of krypton 86 radiation. Science had finally found its natural standard, albeit an abstruse and not readily replicable one. This new meter and the bureau's other changes had no effect on the average person, with the exception of Centigrade being renamed Celsius in 1948. When the base SI measures were set, however, Celsius would not be among them.

Defining essential base units was a big part of what the switch to SI was all about. Over the years, scientists had created informal metric units, with one of the earliest and most popular being the calorie, which was the amount of heat necessary to raise a gram of water 1°C. The calorie was eventually redefined in terms of the joule (the official SI base unit) and its use in science discouraged. A similar measure was "absolute Centigrade," based on an idea by Lord Kelvin to start the temperature scale at absolute zero, thereby eliminating from the thermometer negative numbers, which wreaked havoc on calculations. The measure was to be renamed after Kelvin when it was adopted by the Bureau in 1954, with Celsius becoming a "derived unit" and 0°C redefined as 273.15K. Still, nobody expected people to throw their thermometers away, so degrees Celsius lived on, as did the calorie, now as something like customary metric units.

In his day, Kelvin had pushed harder than anyone else to have Britain adopt the meter, a thing that finally seemed on the verge of happening in 1965, when the UK government announced plans to convert over the next decade. Ever since John Quincy Adams, most everyone had agreed that this was the one event that would pry the United States away from its customary measures, and indeed, the news out of Britain did ignite the American movement. All of a sudden, members of the previously obscure Metric Association were receiving calls to appear before Congress and having their opinions solicited by reporters. The excitement led in 1966 to the launching of the *Metric Today* journal, edited by the group's president, Louis Sokol.

Sokol seemed to see everything through the prism of his cause. During the crushing of the liberal uprising known as Prague Spring, when Sokol (himself of Czech and Slovak descent) fronted the November 1968 *Metric Today* with a note that went: "This issue of the Newsletter has been delayed, because your editor spent several weeks touring Europe. Much of his time was spent in Czechoslovakia where in spite of the brutal Russian occupation he did enjoy using only metric units."

As Congress continued to debate whether or not to approve a comprehensive metric study, Sokol wondered why a report was even needed. America was a century behind the times and should just get on board already. *Metric Today* articles posted dark warnings of what America would be like if it kept resisting the global metric tsunami—even that another Depression would come, one to make the downturn of the 1930s look like a "Sunday picnic."

It was the already bleak economic picture—particularly America's declining exports—that eventually pushed Congress into passing the Metric Study Act of 1968, which approved millions for a massive inquiry by the Bureau of Standards. With legislative action, the meter finally hit the American mainstream. Polls showed that most Americans didn't know what the metric system was, and *Reader's Digest* printed an article that explained the system as a foreign language, with Johnny Carson doing the same on *The Tonight Show*. In 1970, a book called *Prepare Now for a Metric Future* was published, which aimed to both educate the American people and convince them that metric adoption was unavoidable.

The dust jacket of *Prepare* had a geometrically constructivist design that perfectly complemented the subtly threatening imperative of the title. A photo of author Frank Donovan showed an intellectual-looking man with a pointy white beard looking off into the distance, a long pipe attached to the corner of his mouth. Two of Donovan's main themes were introduced on the jacket copy, with a first paragraph beginning "'Inevitable' is the word," and a second one going into how the communists were already metric and exploiting the advantage.

Inside, Donovan quoted the Nobel Prize–winning physical chemist Harold Urey as calling the metric system "a secret weapon of Communism" and claimed that the Russians could never have launched the first satellite if they hadn't been metric (how the United States had just beaten the Soviets to putting a man on the moon was not explained). Donovan implored Americans to end their ostrich act, saying the meter was on its way whether they wanted it or not and writing that "all informed opinion agrees that change is inevitable." To wait was worse than pointless, as the cost would only go up, the way Donovan and other metric advocates brushed off the staggering sums associated with conversion, which the author hypothetically posted at $10 billion over ten years. As it was, the United States was the only industrial nation not committed to metric and was thus relegated to a class with "Botswana, Lesotho, and Western Samoa."

"One reason that Americans supinely accept their illogical system of weights and measures is that they do not know that a better system exists," Donovan wrote. Although sharpest on the backwardness and defective quality of customary measures, Donovan also waxed poetic on the benefits and simplicity of the metric system. He was confident that once they learned it, Americans would love it, and told readers of his book that they could look forward to being asked, "Granddaddy, what was an inch? Mommy says that you would remember."

While many of Donovan's arguments preyed on American fears—of the Soviets, of becoming a third-world country—any fears he and other metric advocates themselves had of what the National Bureau of Standards study would hold for their cause were surely put at ease by the title of the 1971 report: *A Metric America: A Decision Whose Time Has Come.*

Shocking Change

The U.S. Metric Study was the largest research project ever undertaken by the National Bureau of Standards, employing a forty-man team who for three years worked to assemble a report that sought to cover the impact of the metric system on every sector of the American economy, in the process reaching out to hear the opinions of more than seven hundred groups. The final report with supplements ran to more than two thousand pages.

A Metric America presented a plethora of reasons why the United States should and must go metric. Present-day costs, as Donovan had noted, were beside the point in a switch deemed inevitable, while economic benefits would accrue from all sides once metrication was achieved, with increased competitiveness in the global marketplace for American exports and savings at home. The report stated that the United States stood to save as much as half a billion dollars annually from school budgets alone, on the premise that students wasted a full quarter of their time in math class to the eighth grade studying customary measures and those difficult-to-learn nondecimal fractions. (Vulgar fractions, it was believed, were a relic of the past, and unnecessary to be learned at all.)

Although in some respects the report was merely updating nineteenth-century arguments, the way Americans measured things had changed dramatically since then. In the 1800s, prepackaged goods were in their infancy and buying practically anything meant asking for so-and-so much of something and standing there watching it get measured out, a habit pretty much obliterated by self-service supermarkets filled with shelves of premeasured cans, bottles, and boxes, an innovation that dated from the mid-1910s.

An even greater change was education—Americans now knew their decimals. What at the time of Jefferson had been the knowledge of the elite had become the province of grade-schoolers. Decimal fractions were being vigorously taught and kids getting drilled mercilessly in multiplication tables. Reliance on halving and doubling was a thing of the past; no more did politicians and editorialists need worry about the poor illiterate shopkeeper whose livelihood depended on his octal math skills.

But the vital argument that could no longer be made was that America was the world leader in everything and that going metric could only slow the country down; burgeoning inflation and a ballooning debt from the Vietnam War had changed all that. In fact, the economic pressures that had been mounting for a decade came to a head the month after the metric report was released, when the president issued the economic plan known as the Nixon Shock.

On August 15, 1971, Nixon closed the gold window. What this meant was that he pulled America—and therefore the world—off the gold standard. There had been a run on American gold by foreign governments redeeming dollars, and the hemorrhaging stopped only when Nixon forsook Bretton Woods and declared that the dollar was no longer convertible to gold by anybody, or worth any amount of precious metal at all. The dollar would now float against other currencies, each seeking out its own value. This is called *fiat currency*, as bills are worth something only because the government printing them decrees that they are.

When FDR took gold out of the people's hands, currency had ceased being a physical weight; with Nixon, it was no longer a theoretical weight, either. Once upon a time, the measure of land grew or shrank depending on how long it took to work it, whereas money depended upon the cold reality of the scale; now, when other measures were invariable, currency had become a matter purely of opinion. The tombstone of coinage, arguably the most important measure in history, could read:

BORN LYDIA, ANATOLIA, 7TH CENTURY B.C.

DIED WASHINGTON, D.C., 20TH CENTURY A.D.

Now that it was a commodity like anything else, the possession of gold by ordinary Americans was legalized. The price of it skyrocketed, along with inflation. Exacerbating the problem was the Arab oil embargo, which began in October 1973 with drastic rises in the price of gas. This led to gas rationing according to whether your license plate ended in an odd or even number, and hordes of Americans could buy gas only every other day. To save energy, Congress imposed both year-round daylight saving time and a 55-mph speed limit. The American economy turned decrepit and even more intolerably, the ugly saga of Watergate unfolded. When in August 1974 Nixon resigned, it was before an America not just humbled but humiliated.

Against this backdrop, Congress debated how to act on the Bureau of Standards report that stated metric adoption to be a dire national imperative. Should a strict, compulsory deadline for metrication be set, or a voluntary, open-ended push to the new system be encouraged? But while senators and congressmen talked, a most extraordinary thing happened: America was going metric on its own.

A Metric America

Although it would later be remembered for the alleged propensity of its gas tank to explode, the Ford Pinto offered the first metric engines in American-made cars, with 1.6- and 2-liter options. GM's Chevette went the Pinto one better, declaring itself to be a fully metric vehicle. "A new kind of American car," ads announced, "international in design and heritage." The Chevette went on sale in the fall of 1975 and, thanks to its amazing 40 mpg fuel economy, proved a big hit.

Not only in Detroit were big U.S. companies changing. Pharmaceutical manufacturers had begun switching their operations to metric in the 1930s, but prescriptions were starting to follow, with medicine being doled out by the cc (cubic centimeter). Eastman Kodak was another early leader, introducing mm-sized cartridge film in 1963 with the Instamatic, a low-end camera that became a global sensation. Further

American corporate behemoths like IBM, Honeywell, Caterpillar, and International Harvester were planning to go metric as well.

Supermarket shelves started to look different, too, with dual-metric labeling on more than half of packaged goods. The first mainstream American product to be "hard metric" arrived in 1975, when 7UP switched its pints and quarts to half-liter and liter containers, in the process increasing the amount of soda by about 5 percent in each. Its new slogan? *A quart and a liter bit more.*

To be sure, government was helping to facilitate change. Federal education law passed in 1974 made the teaching of the metric system a national policy, forcing school boards and math departments across the country to incorporate it into their curricula. The National Weather Service (the successor to the Weather Bureau) made plans to convert to metric, a move anticipated by weathermen who began giving temperatures in both Celsius and Fahrenheit, while the Federal Highway Administration tested metric road signs. By 1975, more than a dozen states had put up signs like *Chattanooga 25 mi or 40 km* and *Speed Limit 55mph=88km/h.*

California emerged as a metrication leader. A conference organized by the Metric Association drew over a thousand participants to UCLA in 1972. With an eye to the export market, the California wine industry coordinated a voluntary conversion to hard metric bottling, switching from fifths to 0.75-liter bottles. The California Board of Education considered dropping all teaching of customary measures and contracted with a producer to make *When Things Get Hectic, Think Metric,* which joined what had become an onslaught of educational films and books aimed at teaching the metric system.

Imperative titles and the meter went hand in hand, as evidenced by Frank Donovan's *Prepare Now for a Metric Future* (in multiple printings) and his follow-up, *Let's Go Metric,* as well as 1973's *Think Metric!* and 1974's unrelated *Think Metric Now!* A 1974 offering, *You and the Metric System,* was written by an evangelical Christian Goldwaterite aerospace engineer named Allan C. Stover, who had earlier served on Ferdinand Marcos's metrication board in the Philippines.

In the press was general metric cheerleading amid reports of the system's inevitability. On TV and in comic strips, the Peanuts gang struggled with learning the new measures. "By the time we grow up," Peppermint Patty said resignedly, "the metric system will probably be official." An unrelenting litany of miserable puns displayed a supreme lack of creativity on the part of American journalists, who delighted in pointing out how ten-gallon hats would soon measure 37.85 liters, and that McDonald's would need to replace its Quarter Pounder with a Hundred and Thirteen Grammer. Expressions such as touching someone with a ten-foot pole and give an inch/take a yard came in for similar reworkings, while pound cake, foot-long hot dog, and inchworm served as punch lines. More serious laments were expressed over the dulled future literary impact of lines such as Robert Frost's "miles to go before I sleep" and Shakespeare's "Full fathom five thy father lies," as well as the title of Ray Bradbury's *Fahrenheit 451*.

But while writers guffawed over the thought of their favorite football teams facing first down and 9.144 meters, baseball teams began adding meters alongside feet for outfield distances on the walls of such ballparks as Cincinnati's Riverfront Stadium. Wayne, Michigan, declared itself "Metric City USA" and sought to become the first American town to undergo full conversion, while the Blue Ridge Mountain town of Boone, North Carolina, proclaimed itself to be the Kilometer-High City. Denverites could only wring their hands. TAKE ME TO YOUR LITER buttons appeared on lapels, and propaganda even turned sexy. Trading on the bathing-suit poster phenomenon as well as the obsession with female measurements—no Miss America contestant could escape them—a poster was sold that featured a blond-haired beach beauty strapping on her bikini bottom with the slogan *Think Metric* above her and the numbers 92-61-92 below.

"Inevitably and irreversibly," *Time* magazine declared in June 1975, "the metric system is coming to the U.S."

Miles to Go

I am today signing H.R. 8674, the Metric Conversion Act of 1975. This legislation establishes a national policy of coordinating and planning for the increased use of the metric measurement system in the United States. To say that this legislation is historic is an understatement.

And so, just before Christmas, Gerald Ford signed the metric system into U.S. history. The president gave a commemorative pen to the Metric Association—now refashioned as the USMA—but the bill they had lobbied so hard to get was not the one they wanted. No deadlines were imposed; conversion wasn't made compulsory. Not that anyone much noticed. A good portion of the country already believed the ten-year deadline proposed in the 1971 Metric Study was a national fact, and in news reports the bill was treated more as a confirmation than something new.

A U.S. Metric Board was to be appointed by the president to coordinate various aspects of conversion, from education to assisting commerce and industry with the challenges that lay ahead. After all, Ford pointed out, the real impetus to change had come from the private sector. "U.S. industry in this regard," Ford wrote with all the dexterity he was famous for, "is miles ahead of official policy."

That no one on the president's staff picked up a reference to "miles" in a statement on a metric bill was a small indication that things would not go quite as smoothly as expected.

$^{15}/_{16}$

A METRIC AMERICA

Tᴴᴏᴍᴀꜱ Jᴇꜰꜰᴇʀꜱᴏɴ, Benjamin Franklin, John Adams, Lafayette, and George Washington all became celebrities as the United States commemorated the two hundredth year of its existence. The national mascots became a three-man band of two drummers and a head-bandaged flautist, and red, white, and blue covered everything from T-shirts to lunchboxes and fire hydrants. *Schoolhouse Rock!* taught the nation's history on Saturday mornings with cartoon music videos like "The Shot Heard 'Round the World" and "The Preamble," which set the beginning of the constitution to music, turning Gouverneur Morris into a pop lyricist.

The bicentennial celebrations were officially kicked off on April 18, 1975, the anniversary of Paul Revere's ride, when President Ford lit a third lantern at Boston's Old North Church. One if by land, two if by sea, and three for America's third century.

In general, the country bathed itself in optimism, wanting badly to put Vietnam and Watergate in the rearview mirror. By year's end, Ford himself became part of what America was leaving behind, as the people embraced the dark-horse candidate of all time, a peanut-farming former Georgia governor who had studied nuclear physics, Jimmy Carter.

The goodwill generally extended to the metric system, too, as kids bought dual-measure plastic rulers and Datalizer metric conversion slide charts along with their usual school supplies, plus activity books that featured the likes of the cartoon couple Hector and Millie Meter and their dog Killer (Killer Meter—kilometer—get it?). The first official Metric Week kicked off in May 1976, with children using liters instead of cups in Home Ec classes and attending metric school fairs where they got to guess height in meters along with other edifying activities.

Kids, of course, made fun of the whole thing and came up with their own SI units:

$$1000 \text{ mockingbirds} = 1 \text{ kilomockingbird}$$
$$1 \text{ millionth part of a fish} = 1 \text{ microfiche}$$
$$52 \text{ cards} = 1 \text{ decacards}$$

As a new year turned, *Time* proudly announced that it would begin employing metric units in its Science and Medicine sections. "Within the next decade, Americans will be learning a new language in their kitchens, factories, automobiles and local bars." Before January was even out, however, a new poll found Americans feeling rather less jaunty about learning that language.

Mad as Hell

Metric advocates had long been saying that the primary obstacle to metric conversion was a lack of familiarity with the new units, and that getting to know them would breed understanding and acceptance. Early polls backed such beliefs. In 1971, a whopping 56 percent of the population claimed not to even be *aware* of the metric system, and those who held an opinion on converting were split evenly for and against. Two years later, Gallup reported the same percentage of people against the meter, but the group that was for it had gone up by half. In January 1977, however, it was the pro-metric opinion that was flatlining, while opposition to the system had more than doubled.

There were those who considered metric conversion to be unpatriotic. The director of the National Cowboy Hall of Fame in Oklahoma, Dean Krakel, said things like "Metric is definitely communist. One monetary system, one language, one weight and measurement system, one world—all communist." Krakel was a one-man quotation machine, and an ideal face to put to metric opposition—for pro-metric advocates, that is. "We homesteaded 160 acres, not a hectare. We built our homes by board feet. We milked with gallon buckets." Particularly galling to Krakel was a textbook that had Buffalo Bill riding his horse across kilometers of the west.

In 1977, the nationally syndicated columnist Bob Greene began writing articles trumpeting a group he had formed: WAM!, the We Ain't Metric organization. Greene had recently written a book about going on tour with Alice Cooper, where his job had been to dress as Santa Claus and get beaten up. Aside from glam rock, however, Greene would have preferred that American society not progress from the 1950s small Midwestern city where he had grown up. He wrote that he had been stewing "ever since the newspapers and magazines began telling us the wonders of the metric system," and with mounting fulmination—some of it for comedic effect, some of it not—he seethed about American taxpayers footing the bill for metrics to be taught in schools, how ads selling $5 Metricminders claimed that conversion tools would soon be necessary just to get through the supermarket, and that the Arabs— "with some Frenchies and Limeys thrown in"—were to blame.

Rage was in vogue. *Network*'s "I'm as mad as hell, and I'm not going to take this anymore" became a national slogan, and Johnny Paycheck's "Take This Job and Shove It" hit the top of the charts. There was plenty to be mad at, and the metric system became a target for those who thought that watching Johnny Bench and the rest of the Big Red Machine bash home runs over metric outfield walls was just too damn much. But nothing quite got the goat of the American people like the highway signs.

Early results in tests had shown ambivalence among American drivers, which had been taken as encouraging by the Federal Highway Administration, then facing road rage over the recent national switch to a 55 mph speed limit, which had lopped 25 mph off the top speed in some western states. Signs in kilometers quickly became one more thing to hate on the highway, however. For metric advocates, much emphasis had been put on plans to make the nation's road signage metric—even at the $110 million price tag placed on conversion—because it was the one place where the federal government could, by fiat, put the metric system before the people on a daily basis. In April 1977, the FHA announced plans to do just that; the changeover was to begin in July, with an 18-month deadline for all speed limit signs to go metric.

Five thousand letters came in to the federal highway office within weeks, 98 percent of them negative, most of them scathing. "The distress that you are intending to cram down the throats of the American people should be considered treason." "You people must have rocks in your heads." "This is so foolish the ordinary man wonders where your brains are." And from Inez Lawrence of Abilene: "Our people understand our system and are happy with it. Why upset the entire population. WHY?"

In the face of such opposition, President Carter's newly appointed chief of the Federal Highway Administration swiftly axed the plan, making it clear that this was a permanent decision. The move was a blow to the members of the USMA, who heaped blame on the fact that the body that was supposed to steward the nation through metrication—the U.S. Metric Board—had still not been formed, with Ford's selections never having been confirmed by the Senate and Carter's yet to be appointed. The lack of a focused effort was wrecking metrication.

In the national media, some writers were surprised by the resistance but remained confident it was a mere bump in the road. Certainly, all did not appear lost. Coca-Cola pronounced its 7UP experiment a resounding success, finding zero resistance to metric packaging. The ATF extended the hard-metric law from wine bottles to include hard liquor sales. The Chevette was a hit, and GM had gone almost entirely metric, with Ford and Chrysler close behind. The Metric Education Act of 1978 authorized money for grants, and the U.S. Metric Board was finally appointed, confirmed, and set to work. To turn public perception around, the seventeen-member board put out a series of radio PSAs called *Metric Magazine* and hired the *Schoolhouse Rock!* team to create the Metric Marvels, a group of superheroes able to convert yards to meters and pounds to kilograms with stunning ease. Meter Man, Liter Leader, Wonder Gram, and Super Celsius debuted on network TV in the fall of 1978.

But almost as soon as it got started, the Metric Board had to deal with a bomb dropped on the metric cause: a report by the nonpartisan General Accounting Office, sometimes called the Taxpayer's Best Friend. Its findings stood in stark contrast to the massive *A Metric America: A Decision Whose Time Has Come* report from the National Bureau of Standards.

As its first order of business, the GAO strove to make exceedingly clear that—contrary to what everybody seemed to believe—the Metric Conversion Act in no way committed the country to going metric. One of the primary culprits in public misunderstanding was the name of the act itself, which implied that conversion was a national mandate, as well as the actions of certain government agencies. The report also made clear that many of the basic claims regularly put forward by metric advocates had not been proved, especially that U.S. adoption of the metric system was inevitable and that the cost of converting would unquestionably bring about offsetting benefits. It had choice words for what the U.S. Metric Board should do, underlining "The Board is not to advocate metrication." Rather, it should inform the public that national policy did not favor metric over customary measures and hold public hearings on how conversion would affect the American people.

The first recommendation fell largely on deaf ears within the Metric Board, as nearly the entire membership believed its mandate was to bring change, not follow it. But the second one seemed like a good idea, and starting in 1979, the Metric Board undertook on unusual road tour, in which its members got to hear exactly what Americans thought of the system they were pitching.

Kill-A-Meter

You know a state is large and powerful (and thinks a lot of itself) when its governor calls for it to have its own space program. California was not a typical state, however, nor was its governor a typical chief executive. The young and handsome Jerry Brown had been a serious contender for the Democratic presidential nomination in 1976, before Jimmy Carter came out of nowhere and Brown got slapped with the moniker Governor Moonbeam. But Brown was set to challenge Carter again in 1980, although his campaign slogan—"Protect the Earth, serve the people, and explore the universe"—would do him no favors. Nor would a much-lampooned trip to Africa he took with girlfriend Linda Ronstadt that landed the couple on the cover of *Newsweek* in early 1979 looking like a pair of deer in the headlights. As a governor, however, Brown was undeniably innovative, and he was committed to bringing outside voices into government. He had started the California Office of Appropriate Technology, seizing on a movement that took a community-based, environmental approach to technological solutions, and appointed to the office a man Brown had met via a Zen priest: Stewart Brand.

"A thin blond guy with a blazing disk on his forehead" was how Tom Wolfe described Brand in 1968's *The Electric Kool-Aid Acid Test*. "No shirt, however, just an Indian bead necktie on bare skin and a white butcher's coat with medals from the King of Sweden on it." That said, Wolfe noted that Brand belonged to the thoughtful wing of Ken Kesey's Merry Pranksters, a crew he stayed with only fleetingly. Brand could soon be found at Columbia University and other college campuses handing out buttons that asked *Why haven't we seen a photograph of the whole Earth yet?* and claiming that the government was withholding satellite images of the world. Such a picture would foster a sense of universal companionship on the planet, Brand believed, and began publishing *The Whole Earth Catalog*, a compendium of products that became the bible of the back-to-the-land movement as well as the only publication not in book form ever to win the National Book Award.

Brown appointed the maverick Brand to the California Metric Conversion Council, a move blasted by the *Los Angeles Times*, saying it mocked the council's purpose. Brand had made his thoughts on the metric system clear the year before, when he equated it to his other bugbear, nuclear energy. "Both are despicable attempts by government to put an entire people on one 'convenient' system," he wrote. "The one bright idea—nuclear—terrifies people, and the other—metric—infuriates them." His advice for how to fight metric was "Bitch, boycott, and foment."

One of the labor representatives on the U.S. Metric Board, Thomas Hannigan, had himself written antimetric articles prior to his appointment and was trying to get his fellow members to be as neutral as the GAO had recommended. Small-business owners aside, it was organized labor who, fearful of jobs being moved offshore, were proving to be the staunchest opponents of the meter.

A new gasoline crisis triggered by the Iranian revolution provided what other metric board members saw as an opportunity. Gas prices were nearing the $1 per gallon mark, which beyond being a psychological barrier (they began the decade around the 35¢ mark) had triggered a kind of low-tech Y2K-type crisis with gas pumps, which for the most part were unable to register prices past the 99.9¢ mark. Adding dollar-per-gallon capability—one extra digit—would cost about $200 per pump. The USMB floated in public forums the idea of converting the nation's gas stations to liters, thus producing posted prices of around 25¢ and avoiding costly pump alterations.

Amid the worst of the crisis in June 1979, when infuriated Americans sat in miles-long lines to buy gas, the USMB passed a resolution 13 to 1 recommending that fuel pumps go metric, a move with wide industry support. The crisis hit California early—it had the highest gas prices in the continental United States—and thousands of gas stations in the state converted, with Shell and Amoco taking the program nationwide.

Despite Brand's best efforts, California continued to be the most metric-forward state in the union. Even babies in hospitals were getting weighed in kilograms and measured in centimeters, much to the bewilderment of some parents. (Although not those born in Mexico.) In August, the Metric Board road show arrived in the Golden State for "San Francisco Metric Day," an occasion Brand took full advantage of. Speaking at the public forum with his back to members of the board, Brand addressed the audience (and reporters) in a "Stop Metric Madness" T-shirt.

Behind the metric push, Brand claimed, were giant multinational corporations that wanted to be able to sell uniform goods across the world. Why should America give up its measures for their profit? He recognized that the metric system was useful to scientists, but Brand didn't see what improvement it could bring to everyday life, where he saw the customary measures as superior and better suited to human needs. In essence, the meter was a solution to what was not a problem. Whereas advocates had been comparing the metric system to Arabic numerals for over a century, Brand equated it with Esperanto.

Brand also made the point that America wasn't being asked to switch to the official SI measures. What sense did abandoning Fahrenheit for Celsius make when Kelvin was the SI standard? Advocates pointed to Kelvin's being unsuitable for everyday use because it employed such large numbers—72°F was 22°C but 295°K—but that went against a hallowed metric argument, that 30 cm was no less convenient a measure than a foot. No reasonable case could be made for Celsius other than that its adoption would put the whole world on a single system—exactly what Brand abhorred.

Also at the meeting was Brand's nemesis on the California Metric Conversion Council, an engineer named Valerie Antoine. Soon to be president of the Metric Association, Antoine had written several books on SI measures and was as dedicated to the metric cause as anyone alive. She was every bit as feisty as Brand, matching his ostentatious T-shirt with the slide of a dead bird hanging around Uncle Sam's neck—her symbol for the albatross of customary measures. "People talk about clinging to our customary system as if it were some kind of precious tradition," she said, jabbing at Brand. But she cut down anybody who spoke against the meter. Brand, for his part, accused Antoine of wanting to "sneak" metric in; no advocate wanted a public vote on the system, he maintained, because they knew it would lose.

For however much Brand tried to present an intelligent opposition to the metric system, most metric antagonism came purely from the gut. Even though liters presented no major issues for consumers at the pump—after all, drivers didn't generally buy gas in gallons, they bought it by the dollar amount or the tank's worth—Americans were in rebellion against everything metric, and they hated the switch. On Long Island, the Suffolk County legislature banned the sale of gas by the liter in the face of public pressure.

Antimetric forces were having their way with the U.S. Metric Board, too, with Hannigan successfully getting the organization to declare itself officially neutral on the issue of conversion, to the great dismay of Antoine and others. But the greatest indignity was visited upon the pro-metric forces of the board at a public meeting in Phoenix, when members previewed the PSAs they had paid the March of Dimes to produce. They were mortified to discover that the ad mispronounced "kilometer," using the pronunciation modeled on "speedometer" and "thermometer"—ki-LOM-eter—rather than the officially approved KIL-uh-meter.

Foot Ball

How metrication went down in other corners of the Anglosphere provided an alternate universe view to the American experience. Within a short period, nearly every English-speaking country in the world had decided to go metric. Conversion had gone smoothly in South Africa—an authoritarian state that brooked little dissent, as Stewart Brand was fond of pointing out. But Australia had also found metrication a relatively straightforward project. North of the border, Canadians in general parted with their old measures peaceably, and many were caught off guard by the fierceness of the American resistance, causing some Canadians to wonder why they tended to do as they were told.

The UK was a different story altogether. The signature British measurement moment occurred on February 15, 1971—Decimal Day, when the pound sterling abandoned the old £sd 1:20:12 system of Charlemagne and became the world's last major currency to go decimal. D-Day went so smoothly it made metrication seem less daunting, which was a good thing, as one of the conditions of Britain's entry into the European Community (the forerunner to the EU) was a requirement that all products sold in the UK be hard metric as of 1979. By mid-decade, however, the British economic picture had turned as ugly as America's. Anger at the policies of the Labour Party grew, of which conversion to the metric system was one. Not long after taking office in 1979, Prime Minister Margaret Thatcher abolished the ten-year-old Metrication Board, a move that pleased no one so much as John Michell.

In 1970, Michell had self-published a pamphlet called *A Defence of Sacred Measures*, which in look and format seemed more from the era of Thomas Paine than that of *Rolling Stone*. Inside, Michell laid out his antimetric philosophy across a series of brief points. Some of his arguments were similar to those of Brand (who reprinted the piece for American readers in 1978), but his overarching belief was that Anglo-Saxon measures had been handed down to man by a higher power. In some ways Michell came across like a modern-day Charles Piazzi Smyth, and Michell indeed shared some of Piazzi Smyth's ideas, being of nearly like mind when it came to the atheistic and what Michell considered idolatrous origins of the metric system. But his interests ranged wider and farther than Piazzi Smyth's in this respect. Pythagorean mathematics and Plato's theories on harmony and proportion were also important to his support of customary measures, which he believed to be based upon not a pyramid inch, but a megalithic yard.

It is easy to dismiss Michell as a New Age nut when he talks about measures linking mankind to the cosmological order, or to sum him up by his most outrageous adventures, such as taking the Rolling Stones on a field trip to Stonehenge to look for flying saucers. His thinking, however, was more nuanced than it appeared. Part of what Michell was doing was questioning the material culture of the day and supposedly established truth. He and others of like mind styled themselves radical-traditionalists.

One such radical-traditionalist was a young American named Seaver Leslie, who spent a year in Bath assisting Michell on surveys of megalithic monuments. A painter and member of a family of architects from Maine, Leslie was obsessed with the relationship between human proportions in art and architecture, most forcefully exemplified in the writings of Vitruvius and da Vinci's drawing of the Vitruvian man.

Leslie returned to the States in 1978 and was teaching at the Rhode Island School of Design when he threw an antimetric party that, to his surprise, garnered him national attention. In short order, Leslie founded a newsletter called *Footprint* and began hosting events around New York City. In May 1979, Leslie held a Weight and Measure Festival at the Central Park Bandshell, which included a Most Beautiful Foot contest. The prize? A trip to the Great Pyramid.

Further events included a fundraiser at the Dakota, which drew such luminaries as Tom Wolfe, the sculptor Isamu Noguchi, and the designer-photographer Jean-Paul Goude. But Leslie's signature event was the Foot Ball.

Held at the partly abandoned Battery Marine Terminal on the tip of Manhattan, the Foot Ball represented the zenith of the antimetric movement. Leslie took issue with the metric system as "imposed rationality"; such a complaint could never be leveled at his party, attended by eight hundred guests who the *New York Times* said "appeared to be divided between habitués of debutante balls and those who frequent the Mudd Club." Tom Wolfe (who had just begun work on *The Bonfire of the Vanities*) came in his trademark white suit, judged the Most Beautiful Foot contest alongside Kenneth Anger and Charles Rocket, and denigrated the meter as arbitrary and intellectual. Nearly everyone who was anybody in the antimetric movement made an appearance, including Thomas Hannigan, who wished that the other members of the Metric Board could have been there to see it, and John Michell, who had flown in from England to deliver the keynote speech. Leslie, wearing a tuxedo with *Don't Give an Inch* emblazoned on its lapel, had managed to do for the antimetric movement something that the meter had never achieved in its nearly two centuries of existence.

He had made it cool.

Mission Accomplished

Dear Mr. Chairman:

I want to thank the past and present members and staff of the U.S. Metric Board for your service to the Nation in reducing the obstacles to voluntary metrication. You have succeeded in your objective of educating the American people about the meaning of metric measurements in everyday life.

So began the Dear John letter dated March 9, 1982, and signed by the president. Just like that, the metric system became one more corpse in the graveyard of the Reagan budget cuts. It was not, however, a case of partisanship, as when Thatcher nixed the British Metrication Board. The Metric Conversion Act had been signed by a Republican president, and Reagan reiterated his support for the metric system in the same letter in which he killed the board. In fact, denying the metric board its $2.7 million annual funding was probably the only Reagan budget cut that a lifelong liberal Democrat proudly took credit for.

Frank Mankiewicz was no Reagan fan. As Robert F. Kennedy's press secretary in 1967, Mankiewicz had pushed for the senator to debate the then California governor on live national TV. The theme of the telecast was "The Image of America and the Youth of the World," and the politicians were to take questions from students around the globe. Kennedy was not so sure, but Manikewicz believed the brainy senator would cream the lightweight Reagan. He was sorely mistaken, as Reagan proved himself the master of political television and Kennedy came off looking halting and ponderous.

Mankiewicz went on to become an executive at the Peace Corps before taking over the struggling National Public Radio in 1977. Under his tenure, the radio network added a hundred new stations, doubled its audience, and became culturally relevant. NPR's position remained tenuous, however, especially after the Reagan administration slashed its funding by 20 percent. While Mankiewicz worked to make NPR financially independent, the budget ax came down on the Metric Board.

Unknown to virtually anyone, Mankiewicz had the year before sent a letter to his old friend Lyn Nofziger, Reagan's assistant for political affairs, pointing out that the Metric Board was ripe for the cutting. His move had nothing to do with saving money; Mankiewicz was a bona fide antimetricist, and he sent Nofziger a copy of an editorial he had earlier written on the subject. Nofziger was convinced, and he steered Reagan into snuffing out the program.

METRIC: BOARD FADES; AMERICANS BEFUDDLED read a headline in the summer of 1982. One reporter wrote that the killing of any government program, no matter how unpopular, left some sorry to see it go, but he was having an awfully hard time finding mourners for this one.

16/16

or One

ISOLATED

W HETHER THE failure of the 1870s American metric movement as guided by Frederick Barnard was a tragedy is open to debate, but the 1970s attempt surely proved that history repeats itself as farce.

Barnard's cause had been joined by some of the greatest leaders of his era, at a time when measurement reform was all about transcending borders and fueling progress. Such grand goals were long gone by the final quarter of the twentieth century, when metrication in America played out like a petty squabble. Though it's clear in hindsight, no one at the time understood how two revolutions then under way were changing the way the world measured so completely as to make the pro- and antimetric arguments obsolete. One was computing; the other was containerization.

The first full-fledged, fully stacked container ship was the *Gateway City*, launched from Port Newark, New Jersey, on October 4, 1957, the same day as Sputnik. Containerization was the dream of Malcom McLean, the North Carolina–born owner of a large interstate trucking company. Reusable boxes of varying sizes and materials already existed, of course; McLean's innovation was to make the box the vehicle. He did this by separating trailers from the chassis of semi trucks in such a way that they could be reattached to the chassis of trains and turned into railroad cars, or stacked on board ships like giant toy blocks. No more did cargo have to be loaded and unloaded at every change of conveyance. Merchandise previously handled a dozen times might only have to be handled twice. Unloading times at ports plummeted by over 80 percent, causing busy harbors to lose one of their distinctive features, that of long lines of ships waiting to dock. As a vessel made money only while it was moving, ship owners saved a fortune; the time it took to get products to market was slashed, too, saving them even more money by reducing spoilage and the need for warehousing. Sealed containers also virtually eliminated the ancient practice of pilfering by stevedores, a profession which itself all but went extinct.

It has been said that containerization has had a greater impact on global trade than all GATT talks combined. More certainly, it is the single greatest reason stuff is cheap, and it has allowed virtually any product to be sold globally. The container represents a further step in prepackaging, having done for global trade what the supermarket did for retail. In the same way that product no longer had to be measured out for waiting shoppers, the need for measuring bulk goods dwindled, as did the importance of customshouses.

Early on, a new kind of capacity measure was created to deal with the container—the TEU, or twenty-foot-equivalent unit. The TEU was originally set to 8' x 8' x 20', making it a very old-fashioned measure, almost entirely binary and equal to ten cords. Because container sizes vary six inches here and there, it is not an exact standard (making it even more old-fashioned), but the TEU remains the basic measure by which port activity is gauged.

It is deeply ironic that while America was hemming and hawing over whether to go metric on account of its shrinking place in world trade, a U.S. innovation sized in customary feet was becoming the most important measure of capacity ever to hit the global market. Beyond irony, however, containers and TEUs spoke to how unseen the act of measuring had become, and how far removed from the average person. So far removed, in fact, that at a time when more Americans than ever were opposed to metrication—almost two thirds of the country, according to a 1991 poll—and the very thought of it seemed preposterous, the process was accelerating behind the scenes. Conversion of American industry had reached critical mass, and the federal government was finally making a push to go metric, too.

The Omnibus Trade and Competitiveness Act of 1988, signed by Reagan, amended the Metric Act of 1975, considerably strengthening it. It established the metric system as "the preferred system of weights and measures for United States trade and commerce." A nice thought, but what really mattered was that the act mandated all federal agencies go metric by October 1992, except where deemed impractical. The process moved slowly, as many federal agencies found metrication impractical on the face of it, but a 1991 executive order by President Bush put the whip to them, and the most-worked-for goal of Frederick Barnard and his American Metrological Society at long last came to fruition.

The renewed metric push had again been spurred by declining exports and a fear of falling behind, this time to Japan. Among its myriad provisions, the 1988 Omnibus Act also gave fast-track authority to the signing of trade agreements, which put America in a position not only to participate in an era of explosive free trade but to help lead it.

The year 1992 was pivotal, with the signing of NAFTA and a breakthrough in negotiations over the Uruguay Round of the GATT talks, which would result in the creation of the World Trade Organization and the most significant lowering of trade barriers since 1948. The main event of the year, however, took place at Maastricht, with the signing of a document that shared with container units the acronym TEU—the Treaty of European Union.

Metric Martyrs

Official credit for the name "euro" is given to an Esperanto-speaking Belgian, Germain Pirlot, who suggested it in a letter to European Commission president Jacques Santer. The honor should at least be shared with Félix de Parieu, who had in the 1860s tried to deliver the "europe" coin in the hope that it would lead to exactly the sort of thing agreed to at Maastricht 125 years later—a common currency and a European Union.

The euro negotiations had put Great Britain in a bind. The UK generally wanted greater trade with Europe and fewer ties, an arrangement many on the Continent felt was akin to the Brits getting their milk for free without having to buy the cow. In the end, the UK was not alone in having cold feet over the euro and was given an opt-out. Even so, there was no lack of Charles Piazzi Smyths who considered Britain's joining the EU a form of national suicide.

When it came to the meter, there would be no opt-outs. Having been a condition of Britain's joining the EC in 1972, metrication had gone much farther in the United Kingdom than in the United States before stalling during the Thatcher years. This had left the nation in a half imperial, half metric state, a "very British mess" according to the title of one pamphlet denouncing it. The situation would not stand for long, however, as in the wake of Maastricht the hard date for Britain to go metric was set for October 1, 1995.

"M" day, as it was called by some, brought back memories of 1971's "D" day, when the pound sterling went decimal. There was much complaining of the foul Napoleonic wind blowing in from across the Channel, and tales of Savile Row tailors whose ability to tell the weight of fabric by touch was lost forever.

The very few temporary exceptions allowed existed largely in the sentimental realm. Milk was allowed to be sold in pints, but only for returnable bottles. The selling of loose produce was permitted until December 31, 1999, in order to give extra time to those iconic British shopkeepers who needed to acquaint themselves with the mental calculations the change to kilograms would take. But the very idea that on the first day of 2000 it became illegal to sell fruits and vegetables by imperial measures was considered a betrayal of Great Britain by some of those shopkeepers. Like Steve Thoburn.

At the suburban fruit and vegetable stall he kept at the Southwick Market, Thoburn continued to bark "Come and get two pounds of bananas for fifty pence!" in open defiance of the ban. He had already been served two infringement notices when on July 4, 2000, an undercover Consumer Protection Officer patronized Thoburn's stall. Later that same day, two out-in-the-open officers came to Thoburn, demanding to see his scales. The round-faced greengrocer resisted until two policemen intervened, threatening Thoburn with arrest if he didn't comply.

Three confiscated sets of imperial scales, a sack of outlaw bananas, and a swarm of national media later, the affable Thoburn became an English folk hero. He and others like him were soon dubbed the Metric Martyrs. Though at first a tabloid creation, the Metric Martyrs banded together into a formal advocacy group. One protest turned the square outside Parliament into an open-air market where produce was sold by imperial measures only, while a banner outside Thoburn's trial referred to the recent blessed event of the prime minister: "If it's good enough for the Blair baby to weigh in at 6lb 12oz then it is good enough for a haddock."

The Metric Martyr affair took place at a time when globalization's dismantling of traditional ways was being protested far and wide. A trial running more or less concurrently with Thoburn's concerned José Bové, a leftist activist turned sheep farmer living in the Aveyron department of France. Bové had become livid when the World Trade Organization allowed the United States to impose abnormally steep tariffs on high-priced EU imports like Roquefort cheese in retaliation for Europe's banning of hormone-treated American beef. In retaliation for U.S. retaliation, Bové used his tractor to help tear apart a McDonald's that was under construction in the town of Millau.

Millau sat in the heart of the Occitan region of France, which had already suffered much under the weight of forced uniformity from distant political masters. "Occitan" is the name for the language of

Southern France that into the twentieth century had been spoken by a large percentage of the French population. Its usage was all but exterminated in the war against the patois, locally called the Vergonha for the Occitan word meaning shame, which is what the people of the Midi were made to feel for the use of their own tongue.

Bové served a few weeks in jail in 2002 and went on to be elected to the European Parliament. Thoburn's tale ended more sadly with his heart attack in 2004 at age thirty-nine, making him for some a true martyr to metrication. His memory was often conjured when a new wave of protests hit the UK in the face of its final metric exceptions coming due to expire. Technocrats in Brussels, however, finally woke up to the fact that its absolutist approach was causing the EU more bad press than it was worth. The organization's industry commissioner, Günter Verheugen, announced that metrication in Britain had gone far enough.

"I want to bring to an end a bitter, bitter battle that has lasted for decades and which in my view is completely pointless," he said. This didn't amount to any special treatment, as the EU regularly allowed cultural concessions, including for pre-metric relics. The Réaumur temperature scale, for instance, is required in the making of Parmigiano cheese.

Regarding the permanent exceptions granted to Britain, the two main ones had to do with drinking and driving. Miles would remain on road signs, and as Orwell had foreseen, no Englishman would readily stand for his pint being taken away. But the EU was quite correct in deeming UK metrication complete, as having people weigh themselves in stones, hoist pints, and drive miles is a Disneyland kind of diversity. The postmortem on the long, drawn-out conversion effort was written by the *Daily Mail* columnist Peter Hitchens, the more conservative brother of Christopher and most eloquent of England's antimetricists, who lamented: "Perhaps when your culture is taken from you piece by piece, you don't care until it is too late."

Digital Aftermath

With the first bills and coins set to come into use the following year, Europeans could spend 2001 dreaming of how the euro would become the new global currency. At the same time, a last vestige of the coin that had dominated international commerce in the early centuries of the global economy made its exit.

In early 2001, American stock exchanges switched to decimal trading, breaking a tradition that went back to the nation's founding. When the New York Stock Exchange opened, the Philadelphia Mint was still a couple months away from producing the first U.S. coins, and prices were quoted in Spanish pieces of eight. By the dawn of the twenty-first century, however, having stocks listed in quarters, eighths, and sixteenths had become as anachronistic as the use of tally sticks in the early decades of the nineteenth.

Decimalization on Wall Street was one small part of a larger story: the death of the fraction. Although it had been heralded by decimalists going back as far as Simon Stevin, the coup de grâce was delivered by technology. "The fast-growing use of computers has made fractions even more obsolete," NASA had noted decades earlier. The computer had made it easy to type decimal points, but hard to make a fraction. Today, when someone is thinking "one and a half" they will probably type 1.5, while .25 makes a quarter more easily than ¼. It is difficult to imagine old-style fractions on currency, but up to the Civil War, it was common practice. At the time, fractions were more trusted than decimal points; now the opposite is true.

The dying vulgar fraction has taken a toll on customary measures, a trend that began with mechanical counter displays in cash registers and the like. What killed the furlong's eighths if not the odometer's tenths? With digital readouts becoming ubiquitous, the process only accelerated. Scales in supermarkets began to render partial pounds in decimals, not ounces. This creeping decimalization was causing the links between customary measures to be severed, and all but a few to be relegated to specialized use.

For all that 1970s metric advocates got wrong, this was one thing they saw coming. The NASA quote above came in a report endorsing metrication, a stance that might surprise Tom Wolfe and others who had been holding up the space program as proof that anything could be done with the inch and pound. The fact was, NASA's use of customary units had little to do with choice and much to do with the aerospace industry, which long remained dominated by U.S. manufacturers and alone among major American industries was able to resist metrication. It led to NASA's using a mix of metric and customary units, which in turn led to the Mars Orbiter fiasco.

In 1999, the Mars Climate Orbiter disintegrated upon entering the red planet's atmosphere owing to a miscommunication between the spacecraft's software, which used SI newtons, and the ground crew, using pound-force units. The mistake cost NASA—and the American taxpayer—hundreds of millions of dollars, and it provided the A-HA! moment metric supporters had been waiting for.

> Repent! Repent! Surely you now see the error of your ways! Were it not for the continued use in some quarters of English units, encouraged by those such as yourself who make flabby arguments for "diversity," the Mars Climate Orbiter would still be on course. The work of a talented and dedicated group of space engineers has gone down the drain, all because you and others of your ilk perversely enjoy using both feet and meters.

This letter appeared in the *American Journal of Physics* and was addressed to editor Robert H. Romer, who had earlier written a piece ridiculing those in "the SI police" who chastised any appearance of nonmetric units in scientific publications. Romer noted how, when teaching, he would use such units as the jelly donut (10^6 joules of energy), his point being that it was more important for science to be understood than pure. His editorial had brought a wave of uniformly positive support from other physicists—at least until the above missive.

Other than that NASA should be fully metric, lessons to be gleaned from the Mars Orbiter incident are few, especially considering how—far from creating accidents—computer software has done more than anything to allow people to live in a dual-measurement world.

Though they made fractions more difficult, computers have made just about everything else to do with bridging the customary-metric divide a breeze. Conversion is now as easy as speaking "seven ounces to grams" into your smartphone and immediately receiving the answer 198.446662g. We live increasingly in the cyber realm, whereas the zealous metric reformers of Barnard's day inhabited the Industrial Age, which was about nothing so much as the making of physical stuff. The Dewey Decimal System with its hierarchical organization of all human thought belongs to a time when information rested on shelves; its metric-sized card catalogs in their carpentered wooden cases are now gone, and no one thinks of Dewey's decimals as the way knowledge is organized in the world.

Why would Americans go metric when computers have done the job for them and they don't even have to know about it? At a doctor's office, nurses record a patient's height and weight in feet and inches and pounds and ounces, which a software program converts to meters and kilograms. It goes well beyond metric, of course. Cellphones automatically reset themselves at daylight saving or when crossing time zones. With a credit card, currency exchange is seamless. People may tweet with simplified spelling (some consolation for Melvil Dui), but otherwise they mostly let predictive text and spellcheck do the work for them. Improving translation software has led to English being called the last lingua franca. By the same token, the metric system is the last universal system of measure, not because it is indomitable but because the need for such a system is gone.

None of this is the same as saying measurement doesn't matter. America lives with a system that is now nearly unique in the world, one that preserves thousands of years of cultural history. But what *is* our system, exactly? To learn, there is no better place to look than your local supermarket.

Emporium of Measures

EU TO BAN SELLING EGGS BY THE DOZEN blared a *Daily Mail* head-line in the summer of 2010, causing another flare-up of British apoplexy over the bureaucrats in Brussels. The story claimed that count measures were to be made illegal, and that the number of eggs in a carton could no longer be put on its labels. Some worried that they'd have to start buying eggs with no earthly idea of how many they were getting. This, however, was not the case. Eggs could still be packaged in whatever numerical grouping was desired, but they—and all other foodstuffs in the EU—had to be priced by the gram or liter.

On the others side of the Atlantic, selling by the count is alive and well. The produce aisles of American supermarkets are filled with heads of lettuce, bundles of parsley, ears of corn, and lemons all being sold by the piece, or at a discount in twos or threes. In the bakery section, bagels and donuts are offered by the half dozen, and loaves of bread are sold by the one, which might not seem worth mentioning except that in the EU they, too, must be sold by the gram. American eggs, of course, continue to be sold by the dozen—except they aren't. Eggs have long been sold in the United States the same way they are in the EU: by weight. If you've ever puzzled over the meaning of small, medium, large, and extra-large eggs, and thought you bought a carton with some under- or oversize specimens, realize that each size specifies the combined weight of twelve eggs.

In general, sales by weight or capacity dominate American supermarkets, but in quite a different fashion from the *hypermarchés* and *Supermärkte* of Europe. Since a 1994 law that mandated dual labeling, most all products sold in the United States carry metric sizes, which are as straightforward as they are in the EU. The customary sizes they cohabit labels with, however, are a varied lot.

Take products sold by weight. Some are packaged in whole pound sizes, especially pantry staples like flour, coffee, and pasta. Most items, though, are sold in odd ounce sizes about which we give little thought. Do you have any idea how many ounces are in a tube of toothpaste?

Canadians in the 1970s didn't, which is why their government chose it as the first product to be metricized. Who could take offense at the change of a measure they didn't know in the first place?

Capacity measures are where things get confusing, particularly with ounces. When a container of ice cream is marked as being sixteen ounces, are those ounces of weight, dry capacity, or wet capacity?

The answer is wet capacity. In truth, you will rarely come across units of dry capacity, apart from farmstand-style pints of berries. Products measured in liquid capacity, on the other hand, are all over the supermarket, and they are often sold in whole sizes. Whereas you won't see the word "pound" on a 16 oz. bag of chips, 32 oz. containers of anything wet—be it dish soap, apple juice, or chicken stock—will usually be marked as a "quart."

We especially like our customary units front and center when it comes to dairy and beer, just like the Brits. Our milk is sold by the quart and half gallon, and our half-and-half by the half pint. For ice cream, size gives an indication of quality, with mass-market stuff coming in tubs of a quart or more while premium brands get packed into pints. As far as beer, until recently you had to buy it in 12-oz. cans or bottles, and the choice of what came inside was just as restricted. The craft brewing movement changed all that, offering more flavorful drinking options that called to be poured in "honest pints" in bars and brewpubs. Increasingly, you can take the pint size home with you, as 16 oz. cans have come to market, or you can fill a growler, a vessel revived by a Wyoming brewery in 1989 that is usually equal to four pints and has begun to be used for other products as well.

Apart from beer, all other alcohol products must be sold by the liter, making the wine and liquor departments the most metric part of the supermarket. (Assuming, of course, you live in a state that allows such things.) In the soda and water aisles, on the other hand, you come to a section of the store where hard metric and customary sizes live side by side, and marketing decides which product uses which system.

When sold by the liter, carbonated drinks come in big plastic bottles filled with mass-produced drinks full of high-fructose corn syrup. "Artisanal" sodas, on the other hand, are sold in glass bottles, made from cane sugar, and usually sized in ounces. With water, the story is reversed. Here the prestige brands come in liter-sized glass bottles, because the high end of the market has long been dominated by European brands such as Perrier and San Pellegrino.

Most shoppers, of course, couldn't care less. Calories and portion sizes are paid at least as much attention to as grams and ounces, and the most important measure is always in dollars and cents. If we can glean anything from the supermarket, it is that we are a bimensural nation, and then some.

Advocates for the meter got a lot wrong—that the metric system was inevitable, that not adopting it would lead to economic ruin, that it was the secret weapon of our enemies—but where they were most misguided was in thinking that metrication had to be an all-or-nothing proposition.

America has gone metric where it is has been useful to do so. Thankfully, this now includes space travel. Maintaining dual systems in manufacturing, science, and government rarely makes sense. But the fact that it's good to go metric in these areas is not an argument for using it in our daily lives. The metric system can be our operating system without being our interface, and maintaining customary measures will not create a nation of scientific ignoramuses, as has so often been predicted. The great virtue of the metric system is its simplicity; any kid who can't learn it because he or she is forced to also learn the inch and ounce is not a foiled Nobel Prize–winning physicist.

Where converting to metric measures wouldn't be convenient, Americans have kept their customary measures and are glad of it. We are comfortable measuring our bodies in feet and pounds, and we know that inhuman temperatures are those below zero and above 100. The customary measure that causes us the most grief is the month. If we adopted a fixed calendar, we'd no longer miss appointments by mixing up dates and days of the week; if we adopted the kilometer, it would make no practical difference in our lives, except for the immediate aggravation.

But when it comes to how our measures do matter, the important thing about keeping them alive is that they provide an alternate way of thinking. The usefulness of the metric system doesn't change the fact that it is incredibly artificial. Worse, its universality leads to the notion that decimals are the *only* way of perceiving the world.

In the Babylonian sixtieths, Roman twelfths, and medieval halves, quarters, and eighths there is the logic and genius of countless generations of people coming to grasp the world around them, the same way there is logic and genius in the Enlightenment tenths, hundredths, and thousandths of the metric system. What is good about the latter does not negate what is good about the former.

Such arguments are taken as self-evident when it comes to vanishing languages or other living heritages that are endangered. America is preserving ways of thinking that were once common to all humanity, and if we get rid of our measures we will never bring them back. To be for a metric America is to be for a global monoculture.

So how is it that those who cheer José Bové's smashing of a McDonald's and blame the United States for the Coca-Colanization of the planet would want this to happen? How can Americans be stupid, ignorant, and lazy for knowing only one language, and also be those same things for having two systems of measurement? It is because not being metric plays into the idea that America thinks of itself as not having to play by the same rules as the rest of the world. This may be a fair enough criticism in other cases, but not this one.

America has never gone metric because it never had to, and every other country did. Most of them converted while undergoing regime changes, industrializing, and trying to make their people literate and numerate. It used to be that diversity was the enemy of a better life; we now live in an age where the villain has become uniformity.

APPENDIX A:
U.S. CUSTOMARY MEASURES

U.S. CUSTOMARY MEASURES

Unit	Value
Length	
mile	5280 ft
foot	—
inch	1/12th ft
Area	
acre	43,560 sq ft
Volume	
gallon	128 oz
quart	32 oz
pint	16 oz
cup	8 oz
ounce	—
Weight	
ton	1000 lb
pound	—
ounce	1/16th lb

Unlike the Imperial measures (which were set by law) or the Système International measures (which are set by an international organization), customary measures are defined by custom, and thus subject to continual change. The table on the previous page shows the American weights and measures currently in general use. Excluded are units confined to limited or specific use, such as the yard (used mostly with fabrics or on the football field), the fathom (used for sea depth), and all units of dry capacity.

Aside from the cup used in cooking, pints and quarts of fruits and vegetables, and cords of firewood, non-liquid volumes in America are generally given in cubic inches, feet, yards, or meters. The most obvious problem with dry measures is how easily they get confused with wet ones; even though most American kitchens have both a set of measuring cups meant for dry capacity and a large graduated measuring cup meant for liquid capacity, the two types tend to be used indiscriminately. (The difference is significant; two cups of water measured dry is over a quarter-cup more than one measured wet.) Further muddying things are teaspoons and tablespoons, which are neither wet nor dry but *metric*, set to 5ml and 15ml.

The so-called octal math of the Middle Ages is still in modern customary measures. If you double a liquid ounce four times you get a pint; do it seven times and you have a gallon. Inches are divided into halves, quarters, eighths, and so on. This form of math also explains the seemingly inexplicable number of feet in the mile and acre, both of which were based on the rod used in surveying. An acre is 4 × 40 rods (160 sq rods), with 40 rods being a furlong (length of a furrow). Eight furlongs (320 rods) make a mile, and each square mile contains 640 acres. Such numbers—4, 8, 16, 32, 64—made doubling and halving easy; where they got ugly was when survey-based units were related to the foot, as a rod measures 16.5 ft. Thus:

- 1 mile = 320 × 16.5 ft = 5280 ft
- 1 sq rod = 16.5 ft × 16.5 ft = 272.25 sq ft
- 1 acre = 160 × 272.25 sq ft = 43,560 sq ft

For however much such binary math underpins the customary system, in actual use decimals are becoming ever more pervasive. Whereas measures used to be expressed the same way we tell time (e.g., 5:23 = 5/12 hours + 23/60 minutes), we now combine different units only for short distance measures such as 3' 5 1/4". But even this is increasingly expressed as 41.25". Decimal subdivision of units is becoming the rule: The mile, the acre, and the ounces of both weight and capacity are remaindered decimally, as is the pound (except for newborns). For all intents and purposes, American customary measures are now decimal measures.

APPENDIX B:
CUSTOMARY METRIC MEASURES

CUSTOMARY METRIC MEASURES

Unit	SI Value	Equivalency in U.S. Customary Measures
Length		
kilometer	1000m	0.62 miles
meter	—	3.28 ft
centimeter	0.01m	0.39 in
millimeter	0.001m	0.039 in
Area		
hectare	10,000m^2	2.47 acres
Volume		
liter	0.001m^3	1.057 quart
milliliter or cc (cubic centimeter)	0.000001m^3	0.034 oz
Weight		
ton	1,000kg	1.1 U.S. ton
kilogram	—	2.2 lb
gram	0.001kg	0.035 oz
milligram	0.000001kg	0.015 grains

The metric system and the Système International d'Unités (SI) are not the same thing. The SI was born in 1960 and refers to units governed by the BIPM, whereas "metric" is a broad term that can be used for all permutations of the system dating back to the 1790s, as well as for the measures used in everyday life. In SI, all measures of length, area, capacity, and weight are derived from the meter and kilogram, whereas common practice also includes such units as the liter, hectare, and ton. The BIPM recognizes such units as acceptable (unlike the are and calorie), and gives them the symbols l, ha, and t. In theory, all units can be used with all prefixes; in practice, people tend to use only a few permutations of each unit (i.e., centimeters, but not decimeters or decameters), which vary from country to country, or will use a given prefix in only one context, such as measuring the destructive force of bombs in megatons (millions of metric tons of TNT).

ACKNOWLEDGMENTS

I want to thank George Gibson, Jacqueline Johnson, and all the people at Bloomsbury, for publishing not only this book, but so many others that have made this one possible. To Benjamin Adams I owe a particular debt, as he largely set me on this course. A special thank-you goes out to Bess Lovejoy, whose research and advice have been crucial. Then there are those who put up with me over three years of being all too often unavailable, most especially my wife, Andromache Chalfant, and our daughter, Galatea.

NOTES ON SOURCES

Chapter 1/16: THE DAY THE METRIC DIED

The origins and meaning of man-based measures have drawn the speculation of nearly every author on the subject of metrology, but the most
revered treatment on this subject—and much else in the field—is Kula,
Measures and Men.

Chapter 2/16: THOMAS JEFFERSON PLANS

As a resource on American currency, I used Jordan, "Coins of
Colonial and Early America," http://www.coins.nd.edu. The
competing Morris and Jefferson coinage plans are explained in
Garson, "Counting Money: The US Dollar and American
Nationhood, 1781–1820." The best source on decimal coinage is
Tschoegl, "The International Diffusion of an Innovation: The Spread
of Decimal Currency." On Simon Stevin, there is Struik, *The Land of
Stevin and Huygens: A Sketch of Science and Technology in the Dutch
Republic During the Golden Century*; Stevin's work is discussed in
Sarton, "The First Explanation of Decimal Fractions and Measures
(1585). Together with a History of the Decimal Idea and a Facsimile
(No. XVII) of Stevin's *Disme*." For the relation of Jefferson's ideas on

currency, measurement, and division of the Northwest, I was influenced by Linklater, *Measuring America*.

Chapter 3/16: AMERICAN PARIS

On depicting France on the cusp of their revolution, as well as the part of such figures as Benjamin Franklin in it, I was heavily influenced by Schama, *Citizens: A Chronicle of the French Revolution*, while the comment beginning "No-one was more" comes from Siân Evans' translation of *Memoirs of Élisabeth Vigée-Le Brun*. Adams, *The Paris Years of Thomas Jefferson* is an accessible and informative look into a much-explored subject. For the Academy, Condorcet, and Lavoisier, the critical source is Gillispie, *Science and Polity in France: The End of the Old Regime*. On ancien régime measures, the classic study is Kula; see also Zupko, *Revolution in Measurement: Western European Weights and Measures Since the Age of Science*.

Chapter 4/16: METRIC SYSTEMS

Weights and measures in America are discussed in Linklater, *Measuring America*; the same work deals at length with Jefferson's plans, as does Hellman, "Jefferson's Efforts Towards the Decimalization of United States Weights and Measures." On Huygens and the pendulum, see Andriesse, *Huygens: The Man Behind the Principle* and Landes, *Revolution in Time: Clocks and the Making of the Modern World*; for the early Academy, Hahn, *The Anatomy of a Scientific Institution: The Paris Academy of Sciences, 1666–1803*. How the limitations of the seconds pendulum came to be learned is described in Ferreiro, *Measure of the Earth*. The way in which debate over the shape and size of the earth relates to the creation of the meter is described in Alder, *The Measure of All Things: The Seven-Year Odyssey and Hidden Error That Transformed the World*, the single best work on the creation of the metric system.

The translation of Condorcet's comment concerning citizens and calculations I took from Heilbron; for Marat's "*gâteau*" gibe, I looked to Gillispie.

Chapter 5/16: THE DECIMATION OF EVERYTHING

A ripping account of the Terror is found in Schama, *Citizens: A Chronicle of the French Revolution*, while Miller, *Envoy to the Terror: Gouverneur Morris and the French Revolution* utilizes Morris's own words to describe events. On the Academy and weights and measures, see again Gillispie and Hahn for the broad view of the travails of the savants, Alder and Kula for how this relates to weights and measures, and Zupko on the permutations of the early metric system. The fate of the *heure decimal* is explored in Vera, "Decimal Time," and Carrigan, "Decimal Time"; for the French Revolutionary calendar, see Shaw, *Time and the French Revolution: The Republican Calendar, 1789–Year XIV*, and Andrews, *Making the Revolutionary Calendar*. On temperature scales and how Centigrade is not more decimal than Fahrenheit, see Middletown, *A History of the Thermometer and Its Uses in Meteorology*. The translation of David's "Let us destroy . . ." speech I found in Bell's *Lavoisier in the Year One*.

Chapter 6/16: NAPOLEONIC MEASURES

For the history of the Napoleonic period and its enduring ramifications, see Lyons, *Napoleon Bonaparte and the Legacy of the French Revolution* and Woolf, *Napoleon's Integration of Europe*. On Laplace, see Hahn, *Pierre Simon Laplace, 1749–1827: A Determined Scientist*; his conference is described in Crosland, "The Congress on Definitive Metric Standards, 1798–1799: The First International Scientific Conference" and Alder. For the spread of metric standards, see Kula and the various essays in *The Global and the Local: The History of Science and the Cultural Integration of Europe*, in particular Borgato, "The First Applications of the Metric System in Italy." The influence of Beccaria is explored in Maestro, "Going Metric: How It All Started." The spread of the Republican calendar is described in Shaw; of the Napoleonic Code, Limpens, "Territorial Expansion of the Code." As for the reception of the metric system in France, there are two brilliant articles: Alder, "A Revolution to Measure," and Heilbron, "The Measure of Enlightenment," in *Weighing Imponderables and Other Quantitative Science Around 1800*; they are additionally the source for the translated quotations in the chapter, with the exception of "unaccustomed as he was to endless calculations," which is from Kula.

Chapter 7/16: LIGHTHOUSES OF THE SKY

The best source on John Quincy Adams is John Quincy Adams himself. Adams's diary writing began at age twelve and never stopped, running to over 14,000 pages before his death; the Massachusetts Historical Society has much of it scanned and available online. For a general biography, there is Unger, *John Quincy Adams*. Adams's *Report upon Weights and Measures* is vital—if difficult—reading. Treat, *A History of the Metric System Controversy in the United States*, issued by the National Bureau of Standards, delves deeply into the Adams report and is the best single source on weights and measures in the United States to 1971; it also goes into the importance of Hassler, as does Linklater. For the observatory movement, see Portolano, "John Quincy Adams's Rhetorical Crusade for Astronomy," Musto, "A Survey of the American Observatory Movement, 1800–1850," and Paullin, "Early Movements for a National Observatory, 1802–1842."

Chapter 8/16: THE INTERNATIONALISTS

I read Robertson, *Revolutions of 1848* a quarter century ago, and it has ever since colored my idea of the period. For Young America politics, see Eyal, *The Young America Movement and the Transformation of the Democratic Party 1828–1861*. The 1849 Peace Congress is described in the *Proceedings of the Second General Peace Congress* put out by participants, with key aspects discussed in Nichols, "Richard Cobden and the International Peace Congress Movement, 1848–1853." Among the biographical works I consulted for this chapter are: Nash, "Alexandre Vattemare: A 19th-Century Story"; Robb, *Victor Hugo*; Hinde, *Richard Cobden: A Victorian Outsider*; and Donald, *Charles Sumner and the Coming of the Civil War*. On Quetelet and statistics, see Cohen, *The Triumph of Numbers: How Counting Shaped Modern Life*. An underinvestigated piece of history is studied in Randeraad, "The International Statistical Congress (1853–1876): Knowledge Transfers and Their Limits." For the wider context of international congresses see Mazower, *Governing the World*. The story of the International Association is told in Yates, *Narrative of the Origin and Formation of the International Association* and the organization's *Report of the Fourth General Meeting*. The rise of the metric system across Europe is traced in Cox, "The Metric System: A Quarter-Century of Acceptance (1851–1876)."

Chapter 9/16: A Universal Coin

For the tale of the U.S. National Academy, see Cochrane, *The National Academy of Sciences: The First Hundred Years, 1863–1963*, while the organization's own *Proceedings* chronicles its committees and actions. For the various congresses and expositions of the period, I relied much on reports from commissioners and delegates, especially those of Samuel Ruggles, but see Nugent, *Money and American Society, 1865–1880* for the impact of Ruggles himself. The scene of Ruggles in Northampton comes from the 1879 edition of Barnard, *The Metric System of Weights and Measures*. For the postal convention, I looked to Sly, *The Genesis of the Universal Postal Union: A Study in the Beginnings of International Organization*. On the LMU, the search for a universal currency, and the ramifications of both, I was most influenced by Einaudi, *Money and Politics: European Monetary Unification and the International Gold Standard* (1865–1873), but see also the same author's "'The Generous Utopia of Yesterday Can Become the Practical Achievement of Tomorrow': 1000 Years of Monetary Union in Europe" and Reti, *Silver and Gold: The Political Economy of International Monetary Conferences, 1867–1892*. For a contrast between the divergent paths the metric system and universal currency took, see Geyer, "One Language for the World." The story of the creation of an international meter is best told in Quinn, *From Artefacts to Atoms: The BIPM and the Search for Ultimate Measurement Standards*.

Chapter 10/16: THE BATTLE OF THE STANDARDS

On Davies vs. Barnard, see three books of all-but-the-same title: Davies, *The Metric System*; Thomson, *The Metric System: Its Claims as an International Standard of Metrology*; Barnard, *The Metric System of Weights and Measures*. For the Great Pyramid and its impact on measurement reform, the essential work is Reisenauer, "'The Battle of the Standards': Great Pyramid Metrology and British Identity, 1859–1890," but see also Schaffer, "Metrology, Metrication, and Victorian Values," Cox, "The International Institute: First Organized Opposition to the Metric System," and Boorstin, "Afterlives of the Pyramids." The most evenhanded look at Charles Piazzi Smyth's pyramidology comes in Brück and Brück, *The Peripatetic Astronomer: The Life of Charles Piazzi Smyth*; for biographies of metric advocates, I used Chute, *Damn Yankee! The First Career of Frederick A. P. Barnard* and Wiegand, *Irrepressible Reformer: A Biography of Melvil Dewey*. The *Proceedings of the American Metrological Society* tells its own tale; for other views of the organization, see Treat, *A History of the Metric System Controversy in the United States* and Duncan, *Notes on Social Measurement*. The failed first introduction of the meter in Brazil is described in Barman, "The Brazilian Peasantry Reexamined: The Implications of the Quebra-Quilo Revolt, 1874–1875."

Chapter 11/16: STANDARD TIME

For Standard Time and the Meridian Conference, I looked to both Bartky, *One Time Fits All: The Campaigns for Global Uniformity*, which is essential for dispelling certain myths, and Galison, *Einstein's Clocks, Poincaré's Maps*, which is brilliantly written. For Standard Time specifically, see Bartky, "The Adoption of Standard Time." For Abbe, see his own "Report of Committee on Standard Time," and on Abbe himself, Hetherington, "Cleveland Abbe and a View of Science in Mid-Nineteenth-Century America" and Willis and Hooke, "Cleveland Abbe and American Meteorology 1871–1901." In addition to Sandford Fleming's own writings, see Creet, "Sandford Fleming and Universal Time," and for a literary biography, Blaise, *Time Lord: Sir Sandford Fleming and the Creation of Standard Time*. On the transformation of time by the mechanical clock, see Dohrn-van Rossum, *History of the Hour: Clocks and Modern Temporal Orders*, which concludes with the impact of Von Moltke's Reichstag speech. For pyramidologists and the American metric movement, see Cox, "The International Institute: First Organized Opposition to the Metric System."

Chapter 12/16: A TOOLKIT FOR THE WORLD

For the westernization of Japan, I relied on Buruma, *Inventing Japan*. An overview of the 1890s–1920s American metric movement is provided in Treat, *A History of the Metric System Controversy in the United States*. For a more in-depth look at Mendenhall's activities and the role of 1920s metric propagandists, see Vera, "The Social Life of Measures: Metrication in the United States and Mexico, 1789–2004," while the decades-long battle Halsey and Dale waged against Stratton (and each other) is chronicled in Crease, *World in the Balance*. The stories of Volapük and Esperanto are told in Okrent, *In the Land of Invented Languages*. For the connivings of Dewey, see Wiegand, *Irrepressible Reformer*; the role of Carnegie as orthographical activist is told in Anderson, "The Forgotten Crusader: Andrew Carnegie and the Simplified Spelling Movement." A vividly told history of turning back the clock is found in Prerau, *Seize the Daylight: The Curious and Contentious Story of Daylight Saving Time*.

Chapter 13/16: THE GREAT CALENDAR DEBATE

Too little has been written on the movement to fix the calendar. A general overview is given in Cohen, "Adoption and Reform of the Gregorian Calendar," while its origins are described in Rocher, "Fallait-il changer le calendrier en 1884?" What brought about the demise of reform is explored in Davies, Trivizas, and Wolfe, "The Failure of Calendar Reform (1922–1931): Religious Minorities, Businessmen, Scientists, and Bureaucrats." The greatest source of material on calendar reform comes from the writings of Moses Cotsworth and his associates on the one side, and Elisabeth Achelis and her *Journal of Calendar Reform* on the other; both are biased but also reliable. On the 1922 National Calendar Convention and the hearing before the Volstead committee, there is the testimony from the congressional hearing itself. Two important publications from the end of the decade are *The League of Nations and the Reform of the Calendar* (published by the League) and the George Eastman–financed *Report of the National Committee on Calendar Simplification for the United States*.

Chapter 14/16: SHOCKS TO THE SYSTEM

The account of the federal government's nationalization of gold follows Nussbaum, "The Law of the Dollar." The connection between

Sputnik and American interest in metrication is described in Treat, *A History of the Metric System Controversy in the United States*, as is the legislative lead-up to the Metric Study Act, while *A Metric America: A Decision Whose Time Has Come* and the U.S. General Accounting Office, *Getting a Better Understanding of the Metric System—Implications If Adopted by the United States* both provide snapshots of the state of metrication in America. The *Metric Today* newsletter displays the concerns of pro-metric advocates and where their efforts were focused.

Chapter 15/16: A METRIC AMERICA

The best studies of the U.S. attempt to go metric are found in Watkins, "Measures of Change: A Constructionist Analysis of Metrication in the United States" and Watkins and Best, "Successful and Unsuccessful Diffusion of Social Policy: The United States, Canada, and the Metric System." Primary sources can be found on the U.S. Metric Association website, which includes all U.S. metric legislation and period photographs as well as reports on metrication in other countries. I relied on newsweeklies and newspapers to follow the course of metrication and public reaction. For the feelings inside the antimetric movement, I drew upon interviews with Seaver Leslie.

Chapter 16/16: ISOLATED

On the origins and impact of containerization, two sources are Levinson, *The Box: How the Shipping Container Made the World Smaller and the World Economy Bigger* and Cudahy, *Box Boats: How Container Ships Changed the World*. The reasons why the United States retains its customary measures are discussed in Vera, "The Social Life of Measures: Metrication in the United States and Mexico, 1789–2004"; Vera also proves wrong the standard view that the United States, Myanmar, and Liberia are the only three nonmetric countries, adding several Pacific Island nations to the list. A balanced view of America's failure to go metric is given in Alder, *The Measure of All Things: The Seven-Year Odyssey and Hidden Error That Transformed the World*.

BIBLIOGRAPHY

Abbe, Cleveland. "Report of Committee on Standard Time." *Proceedings of the American Metrological Society* II (1880): 17–44.

———. "A Short Account of the Circumstances Attending the Inception of Weather Forecast Work by the United States." *Weather Bureau Topics and Personnel* (April 1916): 1–3.

Achelis, Elisabeth. *"Be Not Silent."* New York: Pageant, 1961.

———. *The Calendar for Everybody.* New York: G. P. Putnam's Sons, 1943.

———. *Of Time and the Calendar.* New York: Hermitage House, 1955.

———. *The World Calendar.* New York: G. P. Putnam's Sons, 1937.

Ackerman, Carl W. *George Eastman.* Boston: Houghton Mifflin, 1930.

Adams, John Quincy. *The Diary of John Quincy Adams.* Edited by Joseph Nevins. New York: Longmans, Green, 1928.

———. *Report upon Weights and Measures.* Washington, D.C.: Gales and Seaton, 1821.

Adams, William Howard. *The Paris Years of Thomas Jefferson.* New Haven: Yale University Press, 1997.

Addison, Paul. "USMA History: John T. Johnson, a True Metric Pioneer Who Was USMA President for 25 Years." *Metric Today* 32, no. 5 (Sept.–Oct. 1997): 4.

Alder, Ken. *The Measure of All Things: The Seven-Year Odyssey and Hidden Error That Transformed the World*. New York: Free Press, 2002.

———. "A Revolution to Measure." In *The Values of Precision*. Edited by M. Norton Wise. 39–71. Princeton: Princeton University Press, 1995.

Alexander, J. H. *International Coinage for Great Britain and the United States*. Oxford: John Henry and James Parker, 1857.

Al-Khalili, Jim. *The House of Wisdom: How Arabic Science Saved Ancient Knowledge and Gave Us the Renaissance*. New York: Penguin, 2011.

American Metrological Society. *Proceedings of the American Metrological Society*, vols. 1–5. New York: American Metrological Society, 1889.

Anderson, George B. "The Forgotten Crusader: Andrew Carnegie and the Simplified Spelling Movement." *Journal of the Simplified Spelling Society* J2 (1999/2): 11–15.

Andrews, George Gordon. "Making the Revolutionary Calendar." *American Historical Review* 36, no. 3 (April 1931): 515–32.

Andriesse, C. D. *Huygens: The Man Behind the Principle*. Cambridge: Cambridge University Press, 2005.

Auerbach, Jeffrey A. *The Great Exhibition of 1851: A Nation on Display*. New Haven: Yale University Press, 1999.

Barman, Roderick J. "The Brazilian Peasantry Reexamined: The Implications of the Quebra-Quilo Revolt, 1874–1875." *Hispanic American Historical Review* 57, no. 3 (Aug. 1977): 401–24.

Barnard, Frederick Augustus Porter. *The Imaginary Metrological System of the Great Pyramid of Gizeh*. New York: John Wiley and Sons, 1884.

———. *The Metric System of Weights and Measures*. New York: Board of Trustees of Columbia College, 1872.

———. "Money." "Weights and Measures." "Longitude and Time." In *Draft Outlines of an International Code*. Edited by David Dudley Field. 281–361. New York: Diossy, 1872.

Bartky, Ian R. *One Time Fits All: The Campaigns for Global Uniformity*. Stanford: Stanford University Press, 2007.

————. "The Adoption of Standard Time." *Technology and Culture* 30, no. 1 (Jan. 1989): 25–56.

Bartlett, David F., ed. *The Metric Debate*. Boulder: Colorado Associated University Press, 1980.

Baxter, W. T. "Early Accounting: The Tally and Checkerboard." *Accounting Historians Journal* 16, no. 2 (Dec. 1989): 43–83.

Bedini, Silvio A. *Thomas Jefferson: Statesman of Science*. New York: Macmillan, 1990.

Bell, Alexander Graham. "Our Heterogeneous System of Weights and Measures: An Explanation of the Reasons Why the United States Should Abandon Its Obsolete System of Inches, Tons, and Gallons." *National Geographic* 27, no. 3 (March 1906): 158–69.

Bell, Madison Smartt. *Lavoisier in the Year One*. New York: W. W. Norton, 2005.

Bernstein, Peter L. *The Power of Gold*. New York: John Wiley and Sons, 2000.

Bigourdan, Guillaume. *Le système métrique des poids et mesures*. Paris: Gauthier-Villars, 1901.

Blackburn, Bonnie, and Leofranc Holford-Strevens. *The Oxford Companion to the Year: An Exploration of Calendar Customs and Time-Reckoning*. New York: Oxford University Press, 1999.

Blaise, Clark. *Time Lord: Sir Sandford Fleming and the Creation of Standard Time*. London: Weidenfeld and Nicolson, 2000.

Boorstin, Daniel J. "Afterlives of the Pyramids." *Wilson Quarterly* 16, no. 3 (Summer 1992): 130–38.

Borgato, Maria Teresa. "The First Applications of the Metric System in Italy." In *The Global and the Local: The History of Science and the Cultural Integration of Europe*. Edited by Michal Kokowski. 438–44. Kraków: Polish Academy of Arts and Sciences, 2007.

Branscomb, Lewis M. "The Metric System in the United States." *Proceedings of the American Philosophical Society* 116, no. 4 (Aug. 15, 1972): 294–300.

Brayer, Elizabeth. *George Eastman: A Biography*. Baltimore: Johns Hopkins University Press, 1996.

Braudel, Fernand. *The Wheels of Commerce (Civilization and Capitalism 15–18th Century, Volume 2)*. Translated by Sian Reynolds. New York: Harper and Row, 1982.

Brooks, Frederick. "The Division of the Day." *Proceedings of the American Metrological Society* 2 (1880): 3–10.

Brück, Hermann Alexander, and Mary T. Brück. *The Peripatetic Astronomer: The Life of Charles Piazzi Smyth*. Bristol: Adam Hilger, 1988.

Bruguière, Michel. "Assignats." In *A Critical Dictionary of the French Revolution*. Edited by François Furet and Mona Ozouf. Translated by Arthur Goldhammer. 426–36. Cambridge, MA: Harvard University Press, 1989.

Buruma, Ian. *Inventing Japan*. New York: Modern Library, 2003.

Buttmann, Günther. *The Shadow of the Telescope: A Biography of John Herschel*. New York: Charles Scribner's Sons, 1970.

Carrigan, Richard A., Jr. "Lessons for the Metric System: Decimal Time." In *The Metric Debate*. Ed. David F. Bartlett. 99–115. Boulder: Associated University Press, 1980.

———. "Decimal Time." *American Scientist* 66, no. 3 (May–June 1978): 305–13.

Chevalier, Michel. *On the Probable Fall in the Value of Gold*. Translated by Richard Cobden. New York: D. Appleton, 1859.

Chute, William J. *Damn Yankee! The First Career of Frederick A. P. Barnard*. Port Washington, NY: Kennikat Press, 1978.

Cochrane, Rexmond C. *The National Academy of Sciences: The First Hundred Years, 1863–1963*. Washington, D.C.: National Academy of Sciences, 1978.

Cohen, Edward L. "Adoption and Reform of the Gregorian Calendar." *Math Horizons* 7, no. 3 (Feb. 2000): 5–11.

Cohen, I. Bernard. *The Triumph of Numbers: How Counting Shaped Modern Life*. New York: Norton, 2005.

Cohen, Victor H. "Charles Sumner and the Trent Affair." *Journal of Southern History* 22, no. 2 (May 1956): 205–19.

Cotsworth, Moses B. *The Rational Almanac*. York: Moses B. Cotsworth, 1902.

Cox, Edward F. "The Metric System: A Quarter-Century of Acceptance (1851–1876)." *Osiris* 13 (1958): 358–79.

———. "The International Institute: First Organized Opposition to the Metric System." *Ohio Historical Quarterly* 68, no. 1 (Jan. 1959): 54–83.

Crease, Robert P. *World in the Balance: The Historic Quest for an Absolute System of Measurement*. New York: W. W. Norton, 2011.

Creet, Mario. "Sandford Fleming and Universal Time." *Scientia Canadensis: Canadian Journal of the History of Science, Technology, and Medicine* 14, no. 1–2 (1990): 66–89.

Crosby, Alfred W. *The Measure of Reality: Quantification and Western Society 1250–1600*. Cambridge: Cambridge University Press, 1997.

Crosland, Maurice. "The Congress on Definitive Metric Standards, 1798–1799: The First International Scientific Conference." *Isis* 60, no. 2 (Summer 1969): 226–31.

Cudahy, Brian J. *Box Boats: How Container Ships Changed the World*. New York: Fordham University Press, 2006.

Dantzig, Tobias. *Number: The Language of Science*. New York: Plume, 2007.

Davenport, Charles B. "Biographical Memoir of Frederick Augustus Porter Barnard." *National Academy of Sciences Biographical Memoirs* 20, no. 10 (1939): 257–72.

Davies, Charles. *The Metric System*. New York and Chicago: A. S. Barnes and Company, 1871.

Davies, Christie, Eugene Trivizas, and Roy Wolfe. "The Failure of Calendar Reform (1922–1931): Religious Minorities, Businessmen, Scientists, and Bureaucrats." *Journal of Historical Sociology* 12, no. 13 (Sept. 1999): 251–70.

Débarbat, Suzanne, and Simone Dumont. "The Decimal Metric System: Facing Population and Scientists in France." In *The Global and the Local: The History of Science and the Cultural Integration of Europe*. Edited by Michal Kokowski. 438–44. Kraków: Polish Academy of Arts and Sciences, 2007.

Devreese, Jozef T. and Guido Vanden Berghe. *'Magic Is No Magic': The Wonderful World of Simon Stevin*. Translated by Lee Preedy. Southampton: WIT Press, 2008.

Dhokalia, Ramaa Prasad. *The Codification of Public International Law*. Manchester: Manchester University Press, 1970.

Dohrn-van Rossum, Gerhard. *History of the Hour: Clocks and Modern Temporal Orders*. Chicago: University of Chicago Press, 1996.

Donald, David Herbert. *Charles Sumner and the Coming of the Civil War*. New York: Knopf, 1960.

———. *Charles Sumner and the Rights of Man*. New York: Knopf, 1970.

Donovan, Frank. *Prepare Now for a Metric Future*. New York: Weybright and Talley, 1970.

Doyle, William. *The Oxford History of the French Revolution*. Oxford: Oxford University Press, 1989.

Duncan, Otis Dudley. *Notes on Social Measurement*. New York: Russell Sage Foundation, 1984.

Einaudi, Luca. *Money and Politics: European Monetary Unification and the International Gold Standard (1865–1873)*. Oxford: Oxford University Press, 2001.

———. "From the Franc to the 'Europe': Great Britain, Germany and the Attempted Transformation of the Latin Monetary Union into a European Monetary Union (1865–73)." *Economic History Review* 53, no. 2 (May 2000): 284–308.

———. "'The Generous Utopia of Yesterday Can Become the Practical Achievement of Tomorrow': 1000 Years of Monetary Union in Europe." *National Institute Economic Review* 172, no. 1 (April 2000): 90–104.

Eyal, Yonatan. *The Young America Movement and the Transformation of the Democratic Party 1828–1861*. New York: Cambridge University Press, 2007.

Ferreiro, Larrie D. *Measure of the Earth: The Enlightenment Expedition That Reshaped Our World*. New York: Basic Books, 2011.

Fleming, Sandford. *Terrestrial Time*. London: Edwin S. Boot, c. 1878.

Fowler, William M., Jr. *American Crisis: George Washington and the Dangerous Two Years After Yorktown, 1781–1783*. New York: Walker, 2011.

Franklin, Benjamin. "Daylight Saving: To the Authors of *The Journal of Paris* [1784]." In *The Ingenious Dr. Franklin: Selected Scientific Letters of Benjamin Franklin*. Edited by Nathan G. Goodman. 17–22. Philadelphia: University of Pennsylvania Press, 1931.

Galison, Peter. *Einstein's Clocks, Poincaré's Maps*. New York: Norton, 2003.

Gandhi, Mohandes K. "Gandhi on Calendar." *Journal of Calendar Reform* 1, no. 4 (Dec. 1931).

Garson, Robert. "Counting Money: The U.S. Dollar and American Nationhood, 1781–1820." *Journal of American Studies* 35, no. 1 (April 2001): 21–46.

Geyer, Martin H. "One Language for the World." In *The Mechanics of Internationalism: Culture, Society, and Politics from the 1840s to the First World War*. Edited by Martin H. Geyer and Johannes Paulmann. 55–92. New York: Oxford University Press, 2001.

Gillispie, Charles C. *Science and Polity in France: The End of the Old Regime*. Princeton: Princeton University Press, 1980.

———. *Science and Polity in France: The Revolutionary and Napoleonic Years*. Princeton: Princeton University Press, 2004.

Hahn, Roger. *The Anatomy of a Scientific Institution: The Paris Academy of Sciences, 1666–1803*. Berkeley: University of California Press, 1971.

———. *Pierre Simon Laplace, 1749–1827: A Determined Scientist*. Cambridge, MA: Harvard University Press, 2005.

Hallock, William, and Herbert T. Wade. *Outlines of the Evolution of Weights and Measures and the Metric System.* New York: Macmillan, 1906.

Halsey, Frederick A., and Samuel S. Dale. *The Metric Fallacy and the Metric Failure in the Textile Industry.* New York: D. Van Nostrand, 1904.

Haynes, Carlyle B. *Calendar Change Threatens Religion.* Washington, D.C.: Religious Liberty Association, 1944.

Heilbron, John L. *Weighing Imponderables and Other Quantitative Science Around 1800.* Berkeley: University of California Press, 1993.

Hellman, C. Doris. "Jefferson's Efforts Towards the Decimalization of United States Weights and Measures." *Isis* 16, no. 2 (Nov. 1931): 266–314.

Hetherington, Norriss S. "Cleveland Abbe and a View of Science in Mid-Nineteenth-Century America." *Annals of Science* 33 (1976): 31–49.

Hinde, Wendy. *Richard Cobden: A Victorian Outsider.* New Haven: Yale University Press, 1987.

Hobsbawm, Eric. *The Age of Capital, 1848–1875.* New York: Charles Scribner's Sons, 1975.

International Association for Obtaining a Uniform Decimal System of Measures, Weights, and Coins. *Report of the Fourth General Meeting.* London: Bell and Dalby, 1860.

International Conference Held at Washington for the Purpose of Fixing a Prime Meridian and a Universal Day. Washington, D.C.: Gibson Bros., 1884.

Isaacson, Walter. *Benjamin Franklin.* New York: Simon & Schuster, 2003.

Ives, Kenneth H. *Written Dialects N Spelling Reforms: History N Alternatives.* Chicago: Progressiv Publishr, 1979.

Jefferson, Thomas. *The Papers of Thomas Jefferson.* Edited by Julian P. Boyd. Princeton: Princeton University Press, 1953.

Jordan, Louis. "The Coins of Colonial and Early America," www.coins. nd.edu.

Kendall, Lane C., and James J. Buckley. *The Business of Shipping*, 7th ed. Centreville, MD: Cornell Maritime Press, 2001.

Kennelly, Arthur E. *Vestiges of Pre-Metric Weights and Measures Persisting in Metric-System Europe, 1926–1927.* New York: Macmillan, 1928.

———. "Proposed Reforms of the Gregorian Calendar." *Proceedings of the American Philosophical Society* 75, no. 1 (May 1935): 71–110.

Kern, Stephen. *The Culture of Time and Space: 1880–1918.* Cambridge, MA: Harvard University Press, 1983.

Klein, Herbert Arthur. *The Science of Measurement.* New York: Dover, 1974.

Kula, Witold. *Measures and Men.* Princeton: Princeton University Press, 1986.

Lampe, Markus. "Explaining Nineteenth-Century Bilateralism: Economic and Political Determinants of the Cobden-Chevalier Network." *Economic History Review* 64, no. 2 (May 2011): 644–68.

Landes, David S. *Revolution in Time: Clocks and the Making of the Modern World.* Cambridge, MA: Belknap Press of Harvard University Press, 1983.

Lawday, David. *Napoleon's Master: A Life of Prince Talleyrand.* New York: St. Martin's Press, 2006.

Leapman, Michael. *The World for a Shilling: How the Great Exhibition of 1851 Shaped a Nation.* London: Headline Book Publishing, 2001.

Lepore, Jill *A Is for American.* New York: Alfred A. Knopf, 2002.

Levinson, Marc. *The Box: How the Shipping Container Made the World Smaller and the World Economy Bigger.* Princeton: Princeton University Press, 2006.

Limpens, Jean. "Territorial Expansion of the Code." In *Code Napoleon and the Common-Law World.* Edited by Bernard Schwartz. 92–109. New York: New York University Press, 1956.

Linklater, Andro. *Measuring America.* New York: Walker and Company, 2002.

Lobingier, Charles Sumner. "Napoleon and His Code." *Harvard Law Review* 32, no. 2 (Dec. 1918): 114–34.

Lyons, Martyn. *Napoleon Bonaparte and the Legacy of the French Revolution*. New York: St. Martin's Press, 1994.

Maestro, Marcello. "Going Metric: How It All Started." *Journal of the History of Ideas* 41, no. 3 (July–Sept. 1980): 479–86.

Major, Ralph H. "Santorio Santorio." *Annals of Medical History* 10, no. 5 (Sept. 1938): 369–81.

Malone, Dumas. *Jefferson and the Rights of Man*. Boston: Little, Brown, 1951.

Mann, William W. "An Historical Popular Description in English and French of the Metrical Decimal System." In *International Exchanges*. Paris: Paul Dupont, 1853.

———. *A New Decimal Metrical System Founded on the Earth's Polar Diameter*. New York and Baltimore: University Publishing Company, 1872.

———. *A New System of Measures, Weights, and Money; Entitled the Linn-Base Decimal System*. New York: University Publishing Company, 1871.

March, Francis A. *The Spelling Reform*. Washington, D.C.: Government Printing Office, 1893.

Marvin, Charles F., and Moses B. Cotsworth. *Moses the Greatest of Calendar Reformers*. Washington, D.C.: International Fixed Calendar League, 1926.

Masayoshi, Matsukata, and Ernest Foxwell. "Report on the Adoption of the Gold Standard in Japan." *Economic Journal* 10, no. 38 (June 1900): 232–45.

Mazower, Mark. *Governing the World*. New York: Penguin, 2012.

McAdie, Alexander. "The Passing of the Fahrenheit Scale." *Geographical Review* 4, no. 3 (Sept. 1917): 214–16.

Mead, Lucia Ames. *A Primer of the Peace Movement*. Boston: American Peace Society, 1904.

Menninger, Karl. *Number Words and Number Symbols: A Cultural History of Numbers*. Translated by Paul Broneer. Cambridge, MA: MIT Press, 1969.

Middleton, W. E. Knowles. *A History of the Thermometer and Its Uses in Meteorology.* Baltimore: Johns Hopkins University Press, 1966.

Miller, Melanie Randolph. *Envoy to the Terror: Gouverneur Morris and the French Revolution.* Washington, D.C.: Potomac Books, 2005.

Morley, John. *The Life of Richard Cobden.* London: T. Fisher Unwin, 1906.

Musto, David F. "A Survey of the American Observatory Movement, 1800–1850." *Vistas in Astronomy* 9, no. 1 (1967): 87–92.

Nash, Suzanne. "Alexandre Vattemare: A 19th-Century Story." *Society of Dix-Neuvièmistes* 3, no. 1 (Sept. 2004): 1–17.

National Academy of Sciences. *Proceedings,* vol. 1. Washington, D.C.: Home Secretary, 1877.

National Bureau of Standards. *A Metric America: A Decision Whose Time Has Come.* Washington, D.C.: National Bureau of Standards, 1971.

Nicastro, Nicholas. *Circumference: Eratosthenes and the Ancient Quest to Measure the Globe.* New York: St. Martin's Press, 2008.

Nichols, David. "Richard Cobden and the International Peace Congress Movement, 1848–1853." *Journal of British Studies* 30, no. 4 (Oct. 1991): 351–76.

Nikolantonakis, Konstantinos. "Weights and Measures: The Greek Efforts to Integrate the Metric System." In *The Global and the Local: The History of Science and the Cultural Integration of Europe.* Edited by Michal Kokowski. 438–44. Kraków: Polish Academy of Arts and Sciences, 2007.

Nugent, Walter T. K. *Money and American Society, 1865–1880.* New York: Free Press, 1968.

Nussbaum, Arthur. "The Law of the Dollar." *Columbia Law Review* 37, no. 7 (Nov. 1937): 1057–91.

O'Brien, Michael J. "Calendar Reform." *Irish Astronomical Journal* 3 (1954): 80–83.

Okrent, Arika. *In the Land of Invented Languages.* New York: Spiegel and Grau, 2009.

O'Mallery, Michael. *Keeping Watch: A History of American Time*. New York: Viking, 1990.

Ostler, Nicholas. *The Last Lingua Franca: English Until the Return of Babel*. New York: Walker, 2010.

Paine, Henry Gallup. *Handbook of Simplified Spelling*. New York: Simplified Spelling Board, 1920.

Parry, J. H. *The Discovery of the Sea*. Berkeley: University of California Press, 1974.

Paullin, Charles O. "Early Movements for a National Observatory, 1802–1842." *Records of the Columbia Historical Society* 25 (1923): 36–56.

Peace Congress Committee. *Proceedings of the Second General Peace Congress*. London: Charles Gilpin, 1849.

Philip, Alexander. *The Reform of the Calendar*. New York: Dutton. 1914.

Piazzi Smyth, Charles. *Life and Work at the Great Pyramid During the Months of January, February, March, and April, A.D. 1865*. Edinburgh: Edmonston and Douglas, 1867.

———. *Our Inheritance in the Great Pyramid*. London: Alexander Strahan, 1864.

Popkin, Jeremy D. *A Short History of the French Revolution*, 3rd ed. Upper Saddle River, NJ: Prentice Hall, 2001.

Portolano, Marlana. "John Quincy Adams's Rhetorical Crusade for Astronomy." *Isis* 91, no. 3 (Sept. 2000): 480–503.

Poulsen, Erling. "Biography of Ole Romer." *Abraham Zelmanov Journal* 1 (2008): 1–6.

Prerau, David. *Seize the Daylight: The Curious and Contentious Story of Daylight Saving Time*. New York: Thunder's Mouth Press, 2005.

Quinn, Terry. *From Artefacts to Atoms: The BIPM and the Search for Ultimate Measurement Standards*. New York: Oxford University Press, 2012.

Randeraad, Nico. "The International Statistical Congress (1853–1876): Knowledge Transfers and Their Limits." *European History Quarterly* 41 (Jan. 2011): 50–65.

Reisenauer, Eric Michael. "'The Battle of the Standards': Great Pyramid Metrology and British Identity, 1859–1890." *Historian* 65 (June 2003): 931–78.

Reppy, Alison, ed. *David Dudley Field: Centenary Essays.* New York: New York University School of Law, 1949.

Reti, Steven P. *Silver and Gold: The Political Economy of International Monetary Conferences, 1867–1892.* Westport: Greenwood Press, 1998.

Richardson, W. F. *Numbering and Measuring in the Classical World.* Exeter: Bristol Phoenix Press, 2004.

Robb, Graham. *Victor Hugo.* New York: W. W. Norton, 1997.

Robertson, Priscilla. *Revolutions of 1848.* Princeton: Princeton University Press, 1952.

Rocher, Patrick. "Fallait-il changer le calendrier en 1884?" Translated by Thomas Dunlap. *L'Astronomie* no. 37 (April 2011): 32–35.

Sarton, George. "The First Explanation of Decimal Fractions and Measures (1585). Together with a History of the Decimal Idea and a Facsimile (No. XVII) of Stevin's *Disme*." *Isis* 23, no. 1 (June 1935): 153–244.

Schama, Simon. *Citizens: A Chronicle of the French Revolution.* New York: Knopf, 1989.

Schaffer, Simon. "Metrology, Metrication, and Victorian Values." In *Victorian Science in Context.* Edited by Bernard V. Lightman. 438–74. Chicago: University of Chicago Press, 1997.

Schiff, Stacy. *A Great Improvisation: Franklin, France, and the Birth of America.* New York: Henry Holt, 2005.

Schiltz, Michael. "Money on the Road to Empire: Japan's Adoption of Gold Monometallism, 1873–97." *Economic History Review* 65, no. 3 (Aug. 2012): 1147–68.

Schimmel, Annemarie. *The Mystery of Numbers.* New York: Oxford University Press, 1993.

Schlesinger, Stephen C. *Act of Creation: The Founding of the United Nations.* Boulder, CO: Westview Press, 2003.

Shalev, Zur. "Measurer of All Things: John Greaves (1602–1652), the Great Pyramid, and Early Modern Metrology." *Journal of the History of Ideas* 63, no. 4 (Oct. 2002): 555–75.

Shaw, Matthew. *Time and the French Revolution: The Republican Calendar, 1789—Year XIV.* Suffolk: Boydell & Brewer, 2011.

Skinner, A. N. "The United States Naval Observatory." *Science* 9, no. 210 (Jan. 6, 1899): 1–16.

Sly, John F. *The Genesis of the Universal Postal Union: A Study in the Beginnings of International Organization.* Worcester; NY: Carnegie Endowment for International Peace, Division of Intercourse and Education, 1927.

Smith, Jeanette C. "Take Me to Your Liter: A History of Metrication in the United States." *Journal of Government Information* 25, no. 5 (Sept.– Oct. 1998): 419–38.

Sobel, Dava. *Longitude.* Walker: New York, 1995.

Stevin, Simon. *Disme: The Art of Tenths, or, Decimall Arithmetike.* Translated by Robert Norton, Gent. London: Hugh Astley, 1608.

Stimson, H. F. "The Present Status of Temperature Scales." *Science* 116, no. 3014 (Oct. 3, 1952): 339–41.

Stover, Allan C. *You and the Metric System.* New York: Dodd, Mead, 1974.

Struik, Dirk Jan. *The Land of Stevin and Huygens: A Sketch of Science and Technology in the Dutch Republic During the Golden Century.* Dordrecht: D. Reidel, 1981.

Sumner, Charles. *Memoir and Letters of Charles Sumner.* Edited by Edward L. Pierce. Boston: Roberts Brothers, 1893.

Taylor, John. *The Battle of the Standards.* London: Longman, Green, Longman, Roberts, and Green, 1864.

Thomson, James B. *The Metric System: Its Claims as an International Standard of Metrology.* New York: Clark and Maynard, 1974.

Treat, Charles F. *A History of the Metric System Controversy in the United States.* Washington, D.C.: National Bureau of Standards, 1971.

Tschoegl, Adrian. "The International Diffusion of an Innovation: The Spread of Decimal Currency." *Journal of Socio-Economics* 39 (2010): 100–109.

Unger, Harlow Giles. *John Quincy Adams*. New York: Da Capo Press, 2012.

————. *Lafayette*. Hoboken: John Wiley and Sons, 2002.

U.S. General Accounting Office. *Getting a Better Understanding of the Metric System—Implications if Adopted by the United States*. Washington, D.C.: Government Printing Office, 1978.

U.S. Metric Board. *Summary Report, July 1982*. Washington, D.C.: Government Printing Office, 1982.

Vattemare, Alexandre. "Letter to the Honorable Hannibal Hamlin." In *International Exchanges*. Paris: Paul Dupont, 1853.

Vera, Hector. "Decimal Time: Misadventures of a Revolutionary Idea, 1793–2008." *KronoScope* 9.1–2 (2009): 29–48.

————. "Melvil Dewey, Metric Apostle." *Metric Today* 45, no. 4 (July–Aug. 2010): 1, 4–6.

————. "The Social Life of Measures: Metrication in the United States and Mexico, 1789–2004." Ph.D. dissertation, New School for Social Research, 2011.

Vigée-Le Brun, Élisabeth. *Memoirs of Élisabeth Vigée-Le Brun*. Translated by Siân Evans. London: Camden, 1989.

Waterston, R. C. *Memoir of George Sumner*. Cambridge, MA: Harvard University Press, 1880.

Watkins, Grace Ellen. "Measures of Change: A Constructionist Analysis of Metrication in the United States." Ph.D. dissertation, Southern Illinois University, 1998.

Watkins, Grace Ellen, and Joel Best. "Successful and Unsuccessful Diffusion of Social Policy: The United States, Canada, and the Metric System." *How Claims Spread: Cross-National Diffusion of Social Problems*. Edited by Joel Best. 267–81. New York: Aldine de Gruyter, 2001.

Wawro, Geoffrey. *The Franco-Prussian War: The German Conquest of France in 1870–1871.* New York: Cambridge University Press, 2003.

Weatherford, Jack. *The History of Money.* New York: Crown, 1997.

Wick, Daniel Lewis. "The Court Nobility and the French Revolution: The Example of the Society of Thirty." *Eighteenth-Century Studies* 13, no. 3 (Spring, 1980): 263–84.

Wiegand, Wayne A. *Irrepressible Reformer: A Biography of Melvil Dewey.* Chicago: American Library Association, 1996.

Willis, Edmund P., and William H. Hooke. "Cleveland Abbe and American Meteorology 1871–1901." *American Meteorological Society* 87, no. 3 (March 2006): 315–26.

Wilford, John Noble. *The Mapmakers.* New York: Alfred A. Knopf, 2000.

Woolf, Stuart. *Napoleon's Integration of Europe.* London: Routledge. 1991.

Yates, James. *Narrative of the Origin and Formation of the International Association.* London: Bell and Daldy, 1856.

Zebel, Sydney H. "Fair Trade: An English Reaction to the Breakdown of the Cobden Treaty System." *Journal of Modern History* 12, no. 2 (June 1940): 161–85.

Zerubavel, Eviatar. "The French Republican Calendar: A Case Study in the Sociology of Time." *American Sociological Review* 42, no. 6 (Dec. 1977): 868–77.

———. *The Seven Day Circle: The History and Meaning of the Week.* New York: Free Press, 1985.

Zupko, Ronald Edward. *Revolution in Measurement: Western European Weights and Measures Since the Age of Science.* Philadelphia: American Philosophical Society, 1990.

IMAGE CREDITS

Richard Cobden, Bibliothèque nationale de France

Lambert Adolphe Jacques Quetelet, Library of Congress

Victor Hugo, author's collection

Elihu Burritt carte de visite, Special Collections, Central Connecticut State University

Charles Sumner, Library of Congress

Samuel Ruggles, copy print of studio portrait, Paris 1863, courtesy of University Archives, Columbia University in the City of New York

Shield nickel back, http://i.colnect.net/

Shield nickel front, http://i.colnect.net/

Dollar franc, liveauctions.holabirdamericana.com

Napoleon III by Alexandre Cabanel, circa 1865, the Walters Art Museum

Frederick A. P. Barnard studio portrait, courtesy of University Archives, Columbia University in the City of New York

Charles Piazzi Smyth, Photo Researchers/SPL/Science Source

Melvil Dewey, Amherst College Archives and Special Collections

Cleveland Abbe, *Popular Science Monthly* volume 32

Sir Sanford Fleming, circa 1865, City of Vancouver Archives

Daylight Saving Time poster, Library of Congress

Moses Cotsworth, Bibliothèque nationale de France

Elisabeth Achelis, Associated Press

"Calendar Change Threatens Religion," author's collection

Tom Wolfe and Seaver Leslie at the May 1981 Foot Ball, photograph by S. Karin Epstein

I-71 Sign Installation in 1979, Ohio Department of Transportation

INDEX

A NOTE ON THE AUTHOR

John Bemelmans Marciano is a *New York Times* bestselling author and illustrator. His work includes the distinctive reference books *Anonyponymous* and *Toponymity* as well as the children's books *Madeline at the White House* and *The 9 Lives of Alexander Baddenfield*. A word and math geek, he lives in Brooklyn with his wife, daughter, and cat.